Computational Techniques for Voltage Stability Assessment and Control

POWER ELECTRONICS AND POWER SYSTEMS

Series Editors
M. A. Pai and Alex Stankovic

Computational Techniques for Voltage Stability Assessment and Control

Venkataramana Ajjarapu
Iowa State University
Ames, Iowa, U.S.A.

 Springer

Venkataramana Ajjarapu
Iowa State University
Department of Electrical and
  Computer Engineering
1122 Coover Hall
Ames, Iowa 50011
U.S.A.

Computational Techniques for Voltage Stability Assessment and Control

Library of Congress Control Number: 2006926216

ISBN-10: 0-387-26080-3   ISBN-10: 0-387-32935-8 (e-book)
ISBN-13: 9780387260808   ISBN-13: 9780387329352 (e-book)

Printed on acid-free paper.

springer.com

TK
1010
A44

Contents

Preface

This book is intended to present bifurcation and continuation based computational techniques for voltage stability assessment and control.

Chapters 1 and 2 provide background material for this book. Chapter 2 reviews various aspects of bifurcation phenomena and includes numerical techniques that can detect the bifurcation points. Chapter 3 discusses the application of continuation methods to power system voltage stability and provides extensive coverage on continuation power flow. Chapter 4 presents general sensitivity techniques available in the literature that includes margin sensitivity. Chapter 5 introduces voltage stability margin boundary tracing. This chapter also discusses application of continuation power flow for ATC. Chapter 6 finally presents time domain techniques that can capture short as well as long term time scales involved in voltage stability. Decoupled time domain simulation is introduced in this chapter. Basic steps involved in various methods in each chapter are first demonstrated through a two bus example for better understanding of these techniques.

I am grateful to Prof. Pai, the series editor, who encouraged me and helped me to write this book.

I would like to acknowledge the help from my previous and current graduate students who helped me directly or indirectly in many ways to organize this book. In general would like to thank Srinivasu Battula, Qin Wang, Zheng Zhou, Gang Shen, Cheng Luo and Ashutosh Tiwari. In particular , I would like to acknowledge contributions of Colin Christy (for chapter 3) , Byongjun Lee (for chapter2) , Bo Long (for chapters 2,3 and 4) , Yuan Zhou (for chapters 3 and 5), Geng Wang (for chapter 5) and Dan Yang (for chapter 6).

I would also like to acknowledge IEEE and Sadhana for some of the figures and material I borrowed from my papers in these journals.

Finally I thank my wife Uma for her continued support and encouragement.

1 Introduction

1.1 What is voltage stability?

Recently IEEE/CIGRE task force [1] proposed various definitions related to power system stability including voltage stability. Fig.1.1 summarizes these definitions.

In general terms, voltage stability is defined as the ability of a power system to maintain steady voltages at all the buses in the system after being subjected to a disturbance from a given initial operating condition. It depends on the ability to maintain/restore equilibrium between load demand and load supply from the power system. Instability that may result appears in the form of a progressive fall or rise of voltages of some buses. A possible outcome of voltage instability is loss of load in an area, or tripping of transmission lines and the other elements by their protection leading to cascading outages that in turn may lead to loss of synchronism of some generators.

This task force further classified the voltage stability into four categories: large disturbance voltage stability, small disturbance voltage stability, short-term voltage stability and long-term voltage stability. A short summary of these classifications is given below.

Large-disturbance voltage stability refers to the system's ability to maintain steady voltages following large disturbances such as system faults, loss of generation, or circuit contingencies. This ability is determined by the system and load characteristics, and the interactions of both continuous and discrete controls and protections. The study period of interest may extend from a few seconds to tens of minutes.

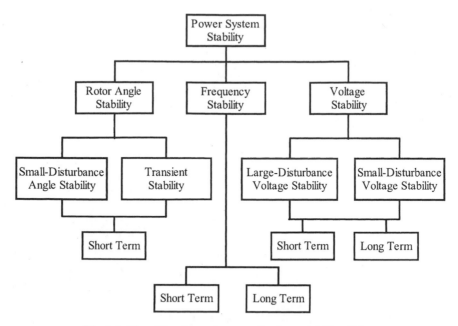

Fig.1.1 Classification of power System stability [1]

Small-disturbance voltage stability refers to the system's ability to maintain steady voltages when subjected to small perturbations such as incremental changes in system load. This form of stability is influenced by the characteristics of loads, continuous controls, and discrete controls at a given instant of time.

Short-term voltage stability involves dynamics of fast acting load components such as induction motors, electronically controlled loads and HVDC converters. The study period of interest is in the order of several seconds, and analysis requires solution of appropriate system differential equations.

Long-term voltage stability involves slower acting equipment such as tap-changing transformers, thermostatically controlled loads and generator current limiters. The study period of interest may extend to several or many minutes, and long-term simulations are required for analysis of system dynamic performance. Instability is due to the loss of long-term equilibrium, post-disturbance steady-state operating point being small-disturbance unstable, or a lack of attraction towards the stable post-disturbance equilibrium. The disturbance could also be a sustained load buildup.

Voltage instability may be caused by various system aspects. Generators, transmission lines and loads are among the most important components.

Generators play an important role for providing enough reactive power support for the power systems. The maximum generator reactive power output is limited by field current limit and armature current limit. Even though reactive power plays an important role in voltage stability, the instability can involve a strong coupling between active and reactive power. When generator reactive capability is constrained by field current limit, the reactive output becomes voltage dependent. The maximum load power is severely reduced when the field current of the local generator becomes limited. Generator limits may also cause limit-induced bifurcation when voltage collapses occur right after the generator limits are reached [2].

Transmission networks are other important constraints for voltage stability. The maximum deliverable power is limited by the transmission network eventually. Power beyond the transmission capacity determined by thermal or stability considerations cannot be delivered.

The third major factor that influences voltage instability is system loads. There are several individual load models due to variety of load devices. Static load models and dynamic load models are two main categories for load modeling. Constant power, constant current and constant impedance load models are representatives of static load models; while dynamic load models are usually represented by differential equations [3]. The common static load models include polynomial or constant impedance, constant current or constant power known as ZIP models. Induction motor is a typical dynamic load model. In real power systems, loads are aggregates of many different devices and thus parameters of load models may be the composite among individual load parameters. Another important load aspect is the Load Tap Changing (LTC) transformer which is one of the key mechanisms in load restoration. During the load recovery process, LTC tends to maintain constant voltage level at the low voltage end. Therefore, load behavior observed at high voltage level is close to constant power which may exacerbate voltage instability.

1.2 Voltage Collapse Incidents

Carson Taylor [4] in his book reported voltage collapse incidents up to the year 1987 (Table 1).   Since then there have been additional incidents that are related to voltage collapse. On July 2nd 1996 the western region (WECC) of the United States experienced voltage collapse. The details of this incident are given in the reference [5]. During May, 1997 the Chilean power system experienced blackout due to voltage collapse that resulted in a loss of 80% of its load. The Chilean power system is mainly radial with prevalent power flows in south north direction. The system configuration is ideal for voltage stability related problems [6].   On July 12, 2004 Athens experienced a voltage collapse that resulted in the blackout of the entire Athens and Peloponnese peninsula [7]. The Hellenic system comprises of generation facilities in the North and West of Greece with most of its load concentrated near the Athens metropolitan region. This system has been prone to voltage stability problems due to the large electrical distance between the generation in the north and load in the Athens region [7]. Greece was then preparing for the Olympic Games that were to be held in Athens. A lot of upgrades and maintenance was scheduled for the system. Unfortunately, most of the planned upgrades were not in place when the event happened. The details of the event and the study are reported by Vournas in the reference [8].

Table 1.1 Voltage Collapse incidents [4]

Date	Location	Time Frame
11/30/86	SE Brazil, Paraguay	2 Seconds
5/17/85	South Florida	4 Seconds
8/22/87	Western Tennessee	10 Seconds
12/27/83	Sweden	50 Seconds
9/22/77	Jacksonville, Florida	Few minutes
9/02/82	Florida	1-3 Minutes
11/26/82	Florida	1-3 Minutes
12/28/82	Florida	1-3 Minutes
12/30/82	Florida	2 Mintues
12/09/65	Brittany, France	?
11/20/76	Brittany, France	?
8/04/82	Belgium	4.5 Minutes
1/12/87	Western France	4-6 Minutes
7/23/87	Tokyo	20 Minutes
12/19/78	France	26 Minutes
8/22/70	Japan	30 Minutes
12/01/87	France	?

There is an extensive literature available covering various aspects of voltage stability. There are excellent text books [4, 9-10], Bulk Power System Voltage Security workshop proceedings [11-16] and IEEE working group publications [17-20] that provide wealth of information that is related to voltage stability. Bibliography related to voltage stability up to 1997 is published in reference [21].   A web based voltage stability search engine is maintained at Iowa state university [22].

The next section provides basic concepts that relate to maximum power transfer through a simple two bus example.

1.3 Two Bus Example

Consider a generator connected to a load bus through a lossless - transmission line as show in Fig.1.2. If   both voltages (E and V) are kept constant then the maximum power transfer occurs at an angle (θ) of 90^0.   The relation between θ and the power transfer (P) through the transmission line is shown in Fig.1.3.

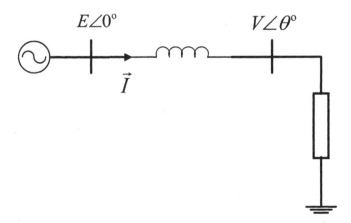

Fig.1.2 A simple two bus system

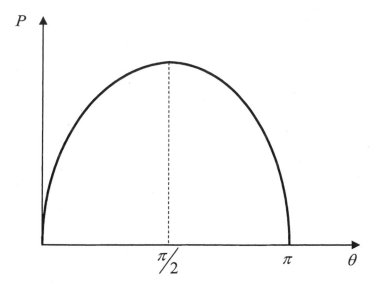

Fig.1.3 The relationship between P and θ

Now consider the same generator with constant terminal voltage being connected to a load bus whose voltage is no longer constant. Then the relation between the load bus voltage and the power transfer through the transmission line is shown in Fig.1.4.

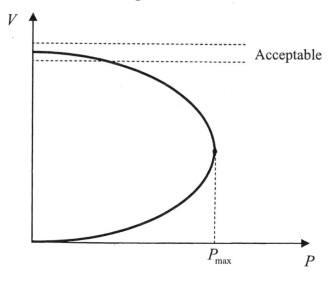

Fig.1.4 Variation load bus voltage with P

With increase in load the voltage at the load bus decreases and reaches a critical value that corresponds to the maximum power transfer.   In general this maximum power transfer is related to voltage instability if the load is constant power type. Beyond this point there is no equilibrium. However if the load is other than constant power then the system can operate below this critical voltage, but draws higher current for the same amount of power transfer.

1.3.1 Derivation for critical voltage and critical power

For this simple example, a closed form solution both for the critical voltage and corresponding maximum power can be derived. From Fig.1.2

$$\vec{V} = \vec{E} - jX\vec{I}$$

$$S = P + jQ = \vec{V}\,\vec{I}^{*} = \vec{V}\left(\frac{\vec{E} - \vec{V}}{jX}\right)^{*} = \vec{V}\left(\frac{\vec{E}^{*} - \vec{V}^{*}}{-jX}\right)$$

$$= V\angle\theta\frac{(E\angle 0 - V\angle - \theta)}{-jX} = -\frac{EV}{X}\sin\theta + j(\frac{EV}{X}\cos\theta - \frac{V^{2}}{X})$$

Separating real and imaginary parts

$$P = -\frac{VE}{X}\sin\theta \tag{1.1}$$

$$Q = \frac{EV}{X}\cos\theta - \frac{V^{2}}{X} \tag{1.2}$$

From Eqs.1.1 and 1.2

$$\sin\theta = -\frac{PX}{EV} \tag{1.3}$$

$$\cos\theta = \left(\frac{QX + V^{2}}{EV}\right)^{2} \tag{1.4}$$

We know: $\sin^{2}\theta + \cos^{2}\theta = 1$ \hfill (1.5)

Use Eq.1.5 to combine Eqs.1.3 and 1.4 into

$$\left(\frac{-PX}{EV}\right)^2 + \left(\frac{QX+V^2}{EV}\right)^2 = 1$$

The above expression can be put into the following form

$$\frac{V^4}{E^4} + \frac{V^2}{E^4}(2QX - E^2) + \frac{X^2}{E^4}(P^2 + Q^2) = 0 \qquad (1.6)$$

Let $v = \dfrac{V}{E}$, $p = \dfrac{PX}{E_2}$, and $q = \dfrac{QX}{E^2}$, then Eq.1.6 becomes (where v, p and q are normalized quantities.)

$$v^4 + v^2(2q-1) + p^2 + q^2 = 0 \qquad (1.7)$$

Let ϕ be the power factor angle of the load, substitute $q = p\tan\phi$ in Eq.1.7 and simplify

$$v^4 + v^2(2p\tan\phi - 1) + p^2\sec^2\phi = 0 \qquad (1.8)$$

Eq.1.8 is quadratic equation in v^2, where

$$v^2 = -(2p\tan\phi - 1) \pm \frac{\sqrt{(2p\tan\phi - 1)^2 - 4p^2\sec^2\phi}}{2} \qquad (1.9)$$

v has four solutions out of which two are physically meaningful. These two physical solutions correspond to high voltage and low voltage solution as shown in Fig.1.5. For example from Eq.1.9, at $p=0$, $v=0$ or 1.

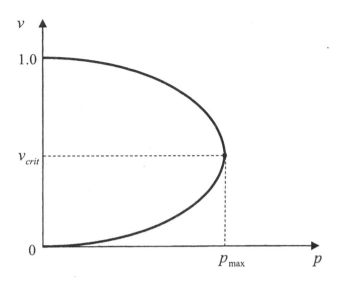

Fig. 1.5   Variation of v with respect to p

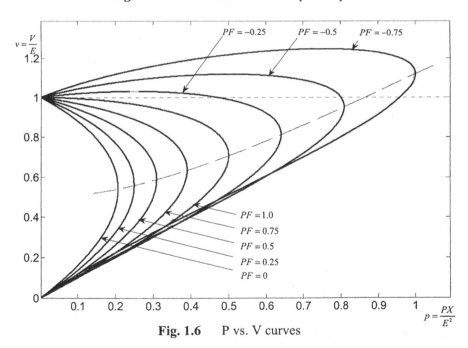

Fig. 1.6   P vs. V curves

At the maximum power point the term inside the square root in equation Eq.1.9 is zero. With this condition, we can show

$$p_{max} = \frac{\cos\phi}{2(1+\sin\phi)} \tag{1.10}$$

$$v_{crit} = \frac{1}{\sqrt{2} * \sqrt{1+\sin\phi}} \tag{1.11}$$

At unity power factor $\phi = 0.0$; $p_{max} = \frac{1}{2} = 0.5$; $v_{crit} = \frac{1}{\sqrt{2}} = 0.707$.

The relationship between θ and ϕ at the maximum power condition can be derived as follows. We know

$$\cos^2\theta = 1 - \sin^2\theta \tag{1.12}$$

From Eq.1.1, at the maximum power conditions, $\sin\theta = -p_{max}/v_{crit}$. Substituting $\sin\theta$ in Eq.1.12 with p_{max} from Eq.1.10 and v_{crit} from Eq.1.11.

$$\cos\theta = \sqrt{1+\sin\phi} = \frac{1}{2v_{crit}} \tag{1.13}$$

Table 1.2 Values of various variables at the critical point

p_{crit}	v_{crit}	q	ϕ	θ
0.5	0.707	0	0	45
0.288	0.577	0.166	30	30
0.207	0.541	0.207	45	22.49
0.1339	0.5175	0.232	60	15
0	0.5	-0.25	90	0
0.866	1	-0.5	-30	60
1.206	1.306	-1.206	-45	67.48
1.86	1.93	-3.22	60	75
.	.	.	.	.
.	.	.	.	.
.	.	.	.	.

1.3.2 Q-V curves

Similar to PV curves one can also obtain QV curves. For each PV curve the power factor is constant, whereas for each QV curve the p is kept constant. From Eq.1.7

$$v^2 = \frac{-(2q-1) \pm \sqrt{(2q-1)^2 - 4(p^2 + q^2)}}{2} \tag{1.14}$$

If we keep p constant in Eq.1.14, then for each p the relation between q and v is shown in Fig.1.7.

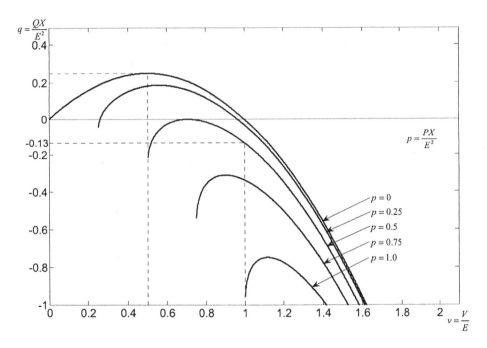

Fig. 1.7   q vs. v curves

We can get q_{crit} by equating the term inside the square root sign to zero in equation Eq.1.14.   Then

$$q_{crit} = \frac{1}{4} - p^2 \tag{1.15}$$

and

$$v_{crit} = \sqrt{\frac{1}{2}(1 - 2q_{crit})}$$

(1.16)

at $p = 0 \rightarrow q_{crit} = 0.25$; $v_{crit} = 0.5$.

Similar to p vs. v curves one can generate q vs. v curves for a given p.   In
the above formulation, we assumed q to be positive for inductive reactive
power.   However, if we assume q as negative for inductive reactive
power, then q vs. v curves can be shown in Fig.1.8. In general in power
system literature q is negative for inductive reactive power.

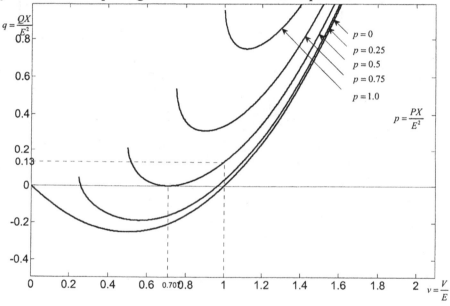

Fig. 1.8 Relationship between voltage and the reactive power

1.3.3 Discussion on PV and QV Curves

PV curves: As mentioned before p vs. v curves can be obtained from
Eq.1.7. These curves are shown in Fig.1.6.   Each curve corresponds to a
particular power factor.   There is a maximum transferable power.   For
any given value of "p" there are two possible voltages (higher voltage with
lower current or lower voltage with higher current).   The normal operation corresponds to high voltage solution.   With capacitor compensation
(leading power factor) the maximum power increases.   However the cor-

responding critical voltage also increases.   From Fig.1.6, one can see that with highly compensated transmission line, normal voltages become critical voltages.

QV curves: These curves give the relation between q and v for a given real power transfer p.   They provide reactive requirement at a given bus to maintain a certain voltage.   For example in Fig.1.7 or Fig.1.8 in the p=0.5curve,to maintain the voltage at 1.0 p.u., a capacitive reactive power injection of q=0.13 p.u. is needed.   If this reactive power injection is lost, the voltage will be decreased to 0.707p.u. which is a critical value (the q= 0 axis just touches the q vs. v curve corresponding to p=0.5). For p=0.5 there is no solution if the net injection is inductive reactive power and this may result in voltage instability. For critical buses, QV curves can be generated from power flow.

Power through transmission lines introduces both real and reactive power loss.   These losses strongly depend on the amount of power through the line.   Transmission lines are mainly dominated by inductive and capacitive characteristics of the line.

At light loads it acts like a capacitor (supply reactive power to the system). At heavy loads it acts like an inductor (absorb reactive power).   The loading at which the inductive and capacitive affects cancel each other is called surge-impedance loading "SIL." "SIL" = approximately 40% to 50% of the line's thermal capacity.

Fig.1.9 [23] shows the relation between line loading and losses.

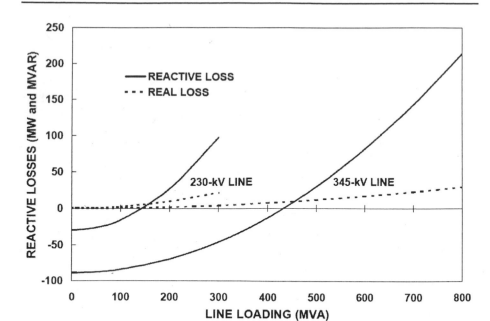

Fig.1.9 Real and reactive power loss vs. line loading for a 100 mile line with the voltages supported at both ends [23]

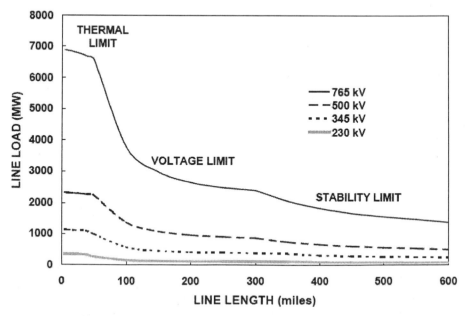

Fig.1.10 Line loading as limited by thermal, voltage and stability [23]

From Fig.1.9 at full line loading reactive losses are 5 times greater than real loss for 230kV line and 9 times greater than real loss for 345kV line.

Fig.1.10 shows the relation between transmission line capacity and the length of the transmission line. Limitations for short, medium and long transmission lines are thermal, voltage drop and stability limits respectively. The above limitations are without any control.

1.3.4 Maximum power and power flow Jacobian

For the two bus example, the power flow equations are:

$$0 = P_0 + \frac{EV}{X}\sin(\theta) = f_1(E,V,\theta)$$

$$0 = Q_0 - \frac{EV}{X}\cos(\theta) + \frac{V^2}{X} = f_2(E,V,\theta)$$

The Jacobian (J):

$$J = \begin{bmatrix} \dfrac{\partial f_1}{\partial \theta} & \dfrac{\partial f_1}{\partial v} \\ \dfrac{\partial f_2}{\partial \theta} & \dfrac{\partial f_2}{\partial v} \end{bmatrix} = \begin{bmatrix} \dfrac{EV}{X}\cos\theta & \dfrac{E}{X}\sin\theta \\ \dfrac{EV}{X}\sin\theta & \dfrac{-E}{X}\cos\theta + \dfrac{2V}{X} \end{bmatrix}$$

The determinant of this Jacobian is $\dfrac{-E^2V}{X^2} + \dfrac{2EV^2}{X^2}\cos\theta$.

Equating this determinant to zero, we get

$$\cos\theta = \frac{1}{2\left(\dfrac{V}{E}\right)} = \frac{1}{2v}$$

This corresponds to the condition of the critical voltage derived in previous section (Eq.1.13).

In general the power system Jacobian becomes singular at the maximum power point. This may leads to convergence problems if one applies the traditional Newton-Raphson method to solve power flow equations.

In this book, computational techniques based on bifurcation and continuation methods will be described to avoid singularities and convergence problems.

References

[1]   Kundur, P., Paserba, J., Ajjarapu, V., Anderson, G., Bose, A., Canizares, C., Hatziargyriou, N., Hill, D., Stankovic, A., Taylor, C. W., Van Cutsem, T., Vittal, V., Definitions and classification of power system stability, IEEE/CIGRE Joint Task force on Stability Terms and Definitions , IEEE transactions on Power Systems, vol. 19, no. 3, pp. 1387-1401, Aug. 2004

[2]   Dobson, I. L., Lu, L., Voltage collapse precipitated by the immediate change in stability when generator reactive power limits are encountered, IEEE Transactions on Circuits and Systems I: Fundamental Theory and Applications, vol. 39, no. 9, pp.762-766, Sept. 1992

[3]   Hill, David J., Nonlinear dynamic load models with recovery or voltage stability studies, IEEE Transactions on Power Systems, vol. 8, no. 1, pp. 166-176, Feb. 1993

[4]   Taylor, C. W., Power system voltage stability, McGraw Hill, 1994

[5]   NERC, 1996 system disturbances: review of selected electric system disturbances in North America, ftp://www.nerc.com /pub/sys/all_up dl/oc/dawg/disturb96.pdf, pp. 22-33, 2002

[6]   Vargas, L., Quintana, V. H., Voltage collapse scenario in the Chilean interconnected system, IEEE Transactions on Power Systems, vol. 14, no. 4, pp. 1415-1421, Nov. 1999

[7]   Vournas, C., Technical summary on the Athens and southern Greece blackout of July 12, 2004

[8]   Vournas, C. D., Manos, G. A., Kabouris, J., Van Cutsem, T., Analysis of a voltage instability incident in the Greek power system, IEEE Power Engineering Society Winter Meeting, vol. 2, 23-27, pp. 1483-1488, Jan. 2000

[9]   Kundur, P., Power system stability and control. New York: McGraw-Hill, 1994

[10]   Van Cutsem, T., Vournas, C., Voltage stability of electric power systems, Kluwer, 1998

[11]   Fink, L. H., Proc. of bulk power system voltage phenomena-I, voltage stability and security, Potosi, MO: ECC/NSF/EPRI Workshop, 1989

[12]   Fink, L. H., Proc. of bulk power system voltage phenomena-II, voltage stability and security, Deep Creek Lake, MD, ECC/NSF Workshop, ECC Inc, 1991

[13]   Fink, L. H., Proc. of bulk power system voltage phenomena-III, voltage stability and security, Davos, Switzerland: ECC/NSF Workshop, 1994

[14] Fink, L. H., Proc. of bulk power system dynamics and control-IV, restructuring, Santorini, Greece: IREP Workshop, 1998

[15] Fink, L. H., Proc. of bulk power system dynamics and control-V, security and reliability in a changing environment, Onomichi, Japan: IREP Workshop, 2001

[16] Proceedings of bulk power system dynamics and control –VI, Managing complexity in power systems: from micro-grids to mega interconnections, Italy, 2004

[17] IEEE Working Group on Voltage Stability, Voltage stability of power systems: concepts, analytical tools, and industry experience, 1990

[18] IEEE Working Group on Voltage Stability, Suggested techniques for voltage stability analysis, 1993

[19] IEEE Power System Relaying Committee Working Group K12, System protection and voltage stability, IEEE publication No. 93, THO 596-7 PWR.

[20] Cañizares, C. A., Editor, Voltage stability assessment: concepts, practices and tools, IEEE-PES Power Systems Stability Subcommittee Special Publication, Product Number SP101PSS (ISBN 0780378695), 2003.

[21] Ajjarapu, V., Lee, B., Bibliography on voltage stability, IEEE Transactions on Power Systems, vol. 13, no. 1, pp. 115-125, Feb. 1998

[22] Iowa State University's Web Based Voltage Stability Search Engine, http://design-2.ece.iastate.edu/biblio/

[23] Kirby, B., Hirst, E., Ancillary service details: voltage control, Oak Ridge National Laboratories/ORNL, http://www.ornl.gov/sci/btc /apps/Restructuring/con453.pdf, Dec. 1997

2 Numerical Bifurcation Techniques

2.1 Various Types of Bifurcation

Nonlinear phenomena relate to the processes that involve physical variables which are governed by nonlinear equations. The models which are described by these equations have been obtained by some approximate projection rationale from presumably more fundamental microscopic dynamics of the system. In some cases a reasonable projection may yield simple linear equations in some approximations.

To demonstrate the basic concepts of nonlinear dynamical systems, we consider a pair of first order coupled ordinary autonomous differential equations. The bases of the classification of these equations are well known and have received much attention in many text books on ordinary differential equations [1, 2].

$$\frac{dx_1}{dt} = f_1(x_1, x_2) \tag{2.1}$$

$$\frac{dx_2}{dt} = f_2(x_1, x_2) \tag{2.2}$$

The equilibrium points are given by $f_1 = 0$ and $f_2 = 0$. Perturb the equilibrium point by Δx_1 and Δx_2, expand the resulting equations in the Taylor Series, and linearize the equations near this equilibrium point. The solutions of Δx_1 and Δx_2 are then given by

$$\Delta x_1 = C_1 e^{\lambda_1 t} + C_2 e^{\lambda_2 t} \tag{2.3}$$

$$\Delta x_2 = C_1 e^{\lambda_1 t} + C_2 e^{\lambda_2 t} \tag{2.4}$$

The constants C_1, C_2, C_3, C_4 are determined by the initial conditions. The exponents λ_1 and λ_2 are the eigenvalues of the Jacobian matrix

$$J = \begin{bmatrix} a & b \\ c & d \end{bmatrix}$$

and can be obtained by solving $|J - \lambda I| = 0$ (where a, b, c, d are the partial derivatives of f_1 and f_2 evaluated w.r.t. x_1 and x_2 at the equilibrium point).

$$\lambda_{1,2} = \frac{1}{2}[Tr(J) + \sqrt{\Delta}]$$

$$Tr(J) = a + d; \ \Delta = \text{discriminant} = Tr(J)^2 - 4\det(J)$$

There are a number of possibilities for the sign and character of λ_1 and λ_2, depending on the signs and relative magnitudes of $Tr(J)$ and $det(J)$. Different possible cases are briefly described below:

Case (i): $Tr(J) < 0$, $\det(J) > 0$, $\Delta > 0$: for these conditions λ_1 and λ_2 are both real and negative. The stationary state is stable and the perturbations decay. It belongs to stable node.

Case (ii): $Tr(J) > 0$, $\det(J) > 0$, $\Delta > 0$: λ_1 and λ_2 are both real and positive. The exponential terms in Eqs.2.3 and 2.4 increase monotonically with time. The perturbations grow exponentially. It belongs to unstable node.

Case (iii): $Tr(J) < 0$, $\det(J) > 0$, $\Delta < 0$: λ_1 and λ_2 are complex and the real part of λ_1 and λ_2 is negative. For this case the perturbations are given by

$$\Delta x = c_1 e^{Re(\lambda t)} \cos(Im(\lambda t) + \theta_1) \qquad (2.3a)$$

$$\Delta y = c_2 e^{Re(\lambda t)} \cos(Im(\lambda t) + \theta_2) \qquad (2.4a)$$

The decaying terms ensure a return to the original stationary state because of the cosine functions. This is a damped oscillatory motion. It belongs to stable focus.

Case (iv): $Tr(J) > 0$, $\det(J) > 0$, $\Delta < 0$: here λ_1 and λ_2 are complex and the real part of λ_1 and λ_2 is positive. The perturbations grow in a divergent oscillatory manner. It is an unstable focus.

Case (v): $Tr(J) > or < 0$, $\det(J) < 0$, $\Delta > 0$: λ_1 and λ_2 are real. $\lambda_1 = +ve$ and $\lambda_2 = -ve$. One of the exponential term in each of Δx_1 and Δx_2 decrease exponentially. The other with the positive root will increase with time. The growing term will eventually dominate and the system will move away from the stationary state. It leads to saddle point behavior.

SPECIAL CASES

   Case (vi): $\det(J) = 0$: here λ_1 and λ_2 are both real.

$$\lambda_1 > 0$$

For $Tr(J) > 0$

$$\lambda_2 = 0$$

$$\lambda_1 = 0$$

For $Tr(J) < 0$

$$\lambda_2 < 0$$

This leads to saddle node bifurcation or fold. To capture the true system behavior, we have to consider nonlinear terms.

   Case (vii): $Tr(J) = 0$, $\det(J) > 0$, $\Delta < 0$: here $\lambda 1$ and $\lambda 2$ are both complex and the real part of these eigenvalues is zero. For this case also, to capture true system behavior, we have to consider the nonlinear terms. This may lead to Hopf bifurcation.

Except for three critical cases: (vi) $det(j) = 0$; (vii) $Tr(J) = 0$; $det(J) > 0$; and a special case where both $det(J) = 0$; $Tr(J) = 0$; the integral curves of the nonlinear system have the same behavior as those of linearized systems in the neighborhood of the equilibrium. These results are summarized with the values of the trace and determinant of the corresponding Jacobian matrix as shown in the phase diagram (Fig.2.1). For linear systems in R^3 [3] make sound classification and arrangement of phase portraits.

However, in the three critical cases mentioned before, the structure of orbits in the state space will change qualitatively. Such a qualitative change in called a bifurcation. This bifurcation may be due to variation of certain parameters in the system. The critical value of the parameter where the bifurcation occurs is the bifurcation value of the parameter.

The chapter is organized as follows: Section 2.2 describes the general principles involved in the study of bifurcation behavior of an n dimensional dynamical system. Sections 2.3, 2.4 and 2.5 discuss the continuation based numerical techniques that can be effectively used to identify various bifurcation points.

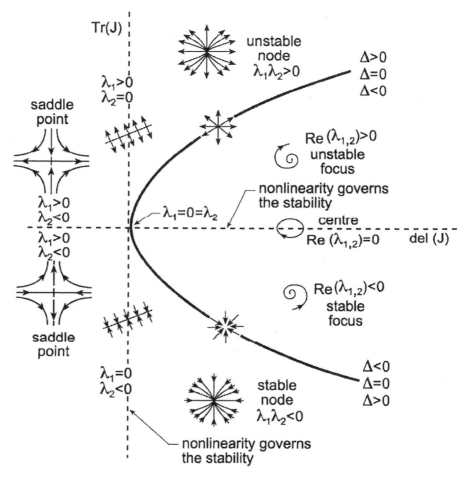

Fig.2.1 Phase diagram [4]

2.2 Bifurcation of Dynamical Systems

Consider a dynamical model of a system [5] described by autonomous differential equations of the vector form in n-dimensional space

$$\dot{x} = F(x, \lambda), x \in R^n, \lambda \in R^k \tag{2.5}$$

Here x denotes the state variables. For power system models these are: generator angles, generator angular velocities, load voltage magnitudes, or angles etc. λ is a vector of time invariant scalar parameters. At an equilibrium point (x_0, λ_0), the left hand term $\dot{x}$ of equation becomes zero, i.e., the steady state solution of Eq.2.5 satisfies the set of nonlinear algebraic equations $F(x_0, \lambda_0) = 0$. If the eigenvalues of the Jacobian $\partial F / \partial x$ become non-zero, then according to implicit function theorem the equilibria of Eq.2.5 can be expressed as the smooth function of $x = x(\lambda)$. The function $x(\lambda)$ is called the branch of equilibria. However if the Jacobian has an eigenvalue with zero real part occurring at some λ, say λ_c, the system $\dot{x} = F(x_c, \lambda_c)$ is structurally unstable and several branches of $x = x(\lambda)$ can come together at (x_c, λ_c) in R^{n+k}. The parameter set λ_c where the system loses its stability is called a bifurcation set. The point (x_c, λ_c) is called bifurcation point. (In general, in engineering systems a one-parameter family with k-1 relations between the parameters $\mu_1, \mu_2, \mu_3, \ldots$ can be represented as a curve, λ, in the k-dimensional parameter space.) Thus the principle of linear stability differentiates between two categories of equilibrium solutions. For the hyperbolic fixed points (where the eigenvalues have non-zero real parts), linear stability analysis suffices completely. For non-hyperbolic fixed points (the points where at least one eigenvalue has zero real part), a linear stability analysis is not applicable and a full nonlinear analysis has to be carried out. There are techniques available to simplify, without any significant loss of information, the representation of the flow in the nonlinear dynamical systems in the neighborhood of non-hyperbolic points. One of these techniques is the center manifold theory. This theory closes the gap left by Hartman-Grobman theorem (HGT). According to HGT, if the Jacobian $\partial F / \partial x$ has no eigenvalues with zero real part, then the family of trajectories near an equilibrium point (x_0, λ_0) of a nonlinear system $\dot{x} = F(x, \lambda)$, and those of the locally linearized system have the same topological structure, which means that in the neighborhood of (x_0, λ_0) there exist homeomorphic mappings which map trajectories of the nonlinear system into trajectories of the linear system. Should, however, an eigenvalue with a zero real part exist, the open question arises how this effects the flow in the neighborhood of the equilibrium point. It is this gap left open by HGT that is closed by the center manifold theory.

2.2.1 Center manifold [6]

Let (x_0, λ_0) be the equilibrium point of $F(x, \lambda)$, and E^s, E^u and E^c the corresponding generalized eigenspaces of the Jacobian matrix $\partial F / \partial x \mid x_0$, where the real part of the eigenvalues (μ) defines the eigenspaces,

$$\text{Re}(\mu) = \begin{cases} < 0 - E^s \\ = 0 - E^c \\ > 0 - E^u \end{cases}$$

Then there exist stable W^s, unstable W^u and center manifold W^c, which are tangential to E^s, E^u, E^c respectively at (x_0, λ_0). If one is interested in the long term behavior (i.e., $t \Rightarrow \infty$) the overall dynamics in the neighborhood of an equilibrium point are reproduced by the flow on the center manifold W^c. This reduction of the dynamics to those in the W^c subspace is the subject of center manifold theory. In order to calculate the flow of the reduced dynamics on W^c, the nonlinear vector field can be transformed to the following form. We can assume that unstable manifold W^u is empty. This makes the presentation simple, without loss of generality.

$$\dot{x}_c = A_c x_c + f(x_c, x_s); X_c \in R^{nc} \tag{2.6}$$

$$\dot{x}_s = A_s x_c + g(x_c, x_s); X_s \in R^{ns} \tag{2.7}$$

The matrix $A_c(n_c, n_c)$ contains n_c, eigenvalues with zero real parts. A_s, matrix (n_s, n_s) contains n_s eigenvalues with negative real parts. The nonlinear functions f and g should be continuously differentiable at least twice and vanish together with their first derivatives at the equilibrium point. X_c correspond to center manifold and are sometimes called active variables. X_s correspond to stable manifold and are called passive variables. Due to nonlinear couplings the influence of x_s in the equation for x_c cannot be ignored. Hence the correct way of analysis is to compute the center manifold.

$$x_s = h(x_c) \tag{2.8}$$

by expressing the dependence of x_s on x_c from Eq.2.8 and then to eliminate from Eq.2.6 to obtain the bifurcation equation

$$\dot{x}_c = A_c x_c + f(x_c, h(x_c)) \tag{2.9}$$

Then the equivalence theorem [5] states that for $t \Rightarrow \infty$, the dynamics of Eq.2.9 in the neighborhood of the equilibrium point is equivalent to the dynamics of the initial system $\dot{x} = F(x, \lambda)$ with λ fixed at the value λ. In order to solve Eq.2.9, one has to know the function $h(x_c)$. This can be obtained as follows

$$\frac{dx_s}{dt} = \frac{dh(x_c)}{dt} = \frac{\partial h}{\partial x_c}\frac{dx}{dt} \tag{2.10}$$

from Eqs.2.6 and 2.7, Eq.2.10 can be written as

$$A_s h(x_c) + g(x_c, h(x_c)) = \left(\frac{\partial h}{\partial x_c}\right)[A_c x_c + f(x_c, h(x_c))] \tag{2.11}$$

or $\left(\dfrac{\partial h}{\partial x_c}\right)[A_c x_c + f(x_c, h(x_c))] - A_s x_c - g(x_c, h_c)) = 0$

The functions h and $(\partial h/\partial x_c)$ are zero at the equilibrium point. Eq.2.11 is in general a partial differential equation which cannot be solved exactly in most cases. But its solution can sometimes be approximated by a series expansion near the equilibrium point. The aforementioned reduction technique of the center manifold theory is similar to its physical counterpart in the slaving principle associated with the synergetic approach proposed by the physicist Herman Haken in the early seventies [7].

In summary, if x is non-hyperbolic then there exist invariant center manifolds tangential to the center subspace and its dimension is equal to the number of eigenvalues of the Jacobian matrix having zero real parts. Then the practically interesting local stability behavior is completely governed by the flow on the center manifold.

Effect of small perturbations of the critical parameters around the bifurcation point can also be studied by unfolding the center manifold. This can be achieved via the method of normal forms [8, 9]. Normal forms play an essential role in bifurcation theory because they provide the simplest system of equations that describe the dynamics of the original system close to the bifurcation points. Even away from the bifurcation point Poincare's theory of normal forms reduces the initial nonlinear equations into the simplest possible forms without distorting the dynamic behavior in the neighborhood of fixed points or periodic solutions. The transformations, which yield to a reduction to normal forms, can be generated by developing the deviations from a state of equilibrium or from periodic motion into power series. Symbolic manipulation packages like MACSYMA, and MAPLE, are helpful in the development of normal forms. Application of normal form away form the bifurcation points to power system examples is given by [10, 11] and examples of the application of center-manifold theory to power systems are given by [12, 13, 14].

The number of possible types of bifurcation increases rapidly with increasing dimension of the parameter space. The bifurcations are organized hierarchically with increasing co-dimension, where co-dimension is the lowest dimension of a parameter space which is necessary to observe a given bifurcation phenomenon. In this book we discuss only the dynamical system with a single parameter variation. Changing this parameter may drive the system into a critical state at which (i) a real eigenvalue becomes zero or (ii) a pair of complex conjugate eigenvalues becomes imaginary. In case (i) new branches of stationary solutions usually arise and are called static bifurcations. (Typical static bifurcations are (i) saddle node or fold, (ii) trans-critical, and (iii) pitchfork.) Case (ii) may lead to the birth of a branch of periodic solutions called dynamic bifurcations. Typical dynamical bifurcation is Hopf.

In many practical engineering problems, identification of these bifurcations is important. For example, buckling load of elastic structures [15] and voltage collapse in power systems [12, 13, 16, and 17] is related to saddle-node bifurcations. Hopf bifurcation and bifurcation of periodic solutions are observed in chemical engineering [18], mechanical engineering [19, 20] and electrical engineering [21, 22, and 50] to name a few. The next section concentrates on the numerical identification of these bifurcations.

2.3 Detection of Bifurcation Points

2.3.1 Static bifurcations

The problem of determining the roots of nonlinear equations is of frequent occurrence in scientific work. Such equations arise typically in connection with equilibrium problems. When describing a real life problem, the nonlinear equations usually involve one or more parameters. Denoting one such parameter by λ, the nonlinear equations read:

$$F(x, \lambda) = 0 \qquad (2.12)$$

where $F: R^n \times R \rightarrow R^n$ is a mapping which is assumed smooth. In Eq.2.12, $\lambda=0$ usually corresponds to the base case solution. If a priori knowledge concerning zero points of F is available, it is advisable to calculate x via a Newton type algorithm defined by an iteration formula such as:

$$x_{i+1} = x_i - A_i^{-1} F(x_i, 0) \qquad i = 0, 1, \dots n \qquad (2.13)$$

where A_i is some reasonable approximation of the Jacobian $F_x(x,0)$. However, if an adequate starting value for a Newton type iteration method is not available, we must seek other remedies. In Section 2.3.2, we will introduce how the lack of knowledge for an initial guess can be tackled by the homotopy method.

Because the systems $F(x, \lambda) = 0$ depends on λ, we speak of a family of nonlinear equations. Solutions now depend on the parameter λ, i.e., $x(\lambda)$. Upon varying the parameter λ, we will get a series of solutions. This is often called a solution curve. At each point corresponding to a certain λ_k, if we keep solving $F(x, \lambda) = 0$ via the conventional Newton type iteration, i.e. by formula (2.13), we may run into difficulty due to the singularity of the Jacobian $F_x(x, \lambda_k)$. The singularity occurs at a so-called turning point (or it is also identified with fold and saddle node) and when the equation is parameterized with respect to λ. In the subsequent sections, we will discuss the interesting topic of curve tracing via the continuations method. We will show how the problem of singularity of the Jacobian can be solved, namely, by switching the continuation parameter.

2.3.2 Homotopy Method

We center our discussion on obtaining a solution to a system of n nonlinear equations in n variables described by Eq.2.12 when λ is at a fixed value. Homotopy method (also some times called embedding method) first defines an easy problem for which a solution is known. Then it defines a path between the easy problem and the problem we actually want to solve. The easy problem, with which the homotopy method starts, is gradually transformed to the solution of the hard problem. Mathematically, this means that one has to define a homotopy or deformation: $R^n \times R \rightarrow R^n$ such that

$$H(x,0) = g(x), H(x,1) = F(x) \qquad (2.14)$$

where g is a trivial smooth map having known zero points and H is also smooth. Typically one may choose a convex homotopy such as

$$H(x,t) = (1-t)g(x) + tF(x) \qquad (2.15)$$

The problem $H(x,t)=0$ is then solved for values of t between 0 and 1. This is equivalent to tracing an implicitly defined curve $c(s) \in H^{-1}(0)$ (i.e. $H(c(s))=0$) for a starting point $(x_0, 0)$ to a solution point $(x_n, 1)$. Under certain conditions, $c(s)$ can be defined as (see Fig2.2):

$$x'(t) = -(H_x(t, x(t)))^{-1} H_t(t, x(t)) \qquad (2.16)$$

If this succeeds, then a zero point of F is obtained, i.e. $H(x,1) = F(x)$. However, the reader may suspect that this is an unnatural approach, since Eq.2.16 seems to be a more complicated problem than to solve $H(c(s))=0$ as a stabilizer. This is the general idea in the continuation methods with a predictor and corrector tracing scheme.

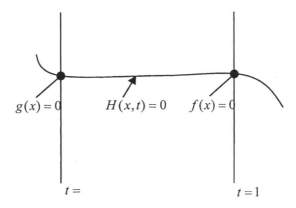

Fig.2.2 Homotopy solution

The relationship Eq.2.15, which embeds the original problem in a family of problems, gives an example of a homotopy that connects the two functions F and g.   In general a homotopy can be any continuous connection between F and g.   If such a map H exists, we say that F is homotropic to g.   A simple two-dimensional nonlinear problem is given here to illustrate how the homotopy method works. The details of this method are given in [23].

Numerical example 1 :[24]

$$F(x) = \begin{bmatrix} f_1(x) \\ f_2(x) \end{bmatrix} = \begin{bmatrix} x_1^2 - 3x_2^2 + 3 \\ x_1 x_2 + 6 \end{bmatrix}$$

$$[X]^T = \begin{bmatrix} x_1 \\ x_2 \end{bmatrix}$$

Define the homotopy function as:

$$H(x,t) = tF(x) + (1-t)g(x)$$
$$= tF(x) + (1-t)F(x) - F(x_0)$$
$$= F(x) + (t-1)F(x)$$

Then we get a curve (from Eq.2.5) defined by:

$$\begin{bmatrix} \dot{x}_1(t) \\ \dot{x}_2(t) \end{bmatrix} = -\frac{1}{\Delta}\begin{bmatrix} x_1 & 6x_2 \\ -x_2 & 2x_1 \end{bmatrix}\begin{bmatrix} 1 \\ 7 \end{bmatrix} = -\frac{1}{\Delta}\begin{bmatrix} x_1 & 42x_2 \\ -x_2 & 14x_1 \end{bmatrix}$$

where $\Delta=2x_1^2+6x_2^2$, with $x_0=(1,1)$. After tracing the implicitly defined curve via some continuation method, we arrive at a solution when $t = 1$: $x^*= (-2.961, 1.978)$.   A real root of F is $(-3, 2)$.   Reasonably we can expect that Newton's method would work well with x^* as the initial guess. After one step of Newton-Raphson iteration, we get $x_1= (-3.0003, 2.0003)$.

However, if we start the Newton's methods directly with the initial guess $x_0= (1, 1)$, it takes more than 5 iterations to get the answer x_1.   For a more complicated practical nonlinear problem, the conventional Newton's method might not work at all due to the poor selection of the initial values.

Whether or not the tracing of a curve can succeed depends on the continuation strategy employed.   If the curve can be parameterized with respect to the parameter t, then the classical embedding algorithm [23] can be applied.   In the following sections, we will discuss how a parameterization is done and how vital this procedure is in the continuation, or say the curve tracing process.   Particularly, we will show how the continuation is carried on even when the curve is not parameterizeable with respect to a certain parameter.

2.3.3 Continuation methods

General description of different aspects of continuation methods with minimum mathematical details in curve tracing is given below. For detailed explanation and mathematical proofs of these methods, please refer to the mathematical references provided in this section. Brief but more pertinent exploration of applying the methodology to power system studies is given in Chapter 3. The system of nonlinear equations in the form of equation Eq.2.12 serves as a basis for discussion. One note to make here is that, for the tracing of a curve defined by Eq.2.15, the discussion is the same as for the curve defined by Eq.2.12.   Here, x denotes an n-dimensional vector.

Continuation methods usually consist of the following [25]: predictor, parameterization strategy, corrector and step length control. Assume that at least one solution of equation Eq.2.12 has been calculated, for instance, by the homotopy method. For the tracing of a curve defined by Eq.2.15, this corresponds to the assumption that g has a known zero point. The jth continuation step starts from a solution (x_{j+1}, λ_j) of Eq.2.12 and attempts to calculate the next solution (x_{j+1}, λ_{j+1}), for the next λ, namely λ_{j+1}. With a predictor-corrector method, the step j to step $j+1$ is split into two parts, with (x_{j+1}, λ_j) produced in between by the prediction. In general, the predictor merely provides an initial guess for the corrector iterations that home in a solution of equations Eq.2.12. The distance between two consecutive solutions is called the step size. In addition to equation Eq.2.12, a relation that identifies the location of a solution on the branch is needed. This identification is closely related to the kind of parameterization strategy chosen to trace the curve.

In the curve tracing process, at some critical points (e.g. turning or fold points), the singularity of the Jacobian matrix F_x often causes trouble either in the prediction or in the correction process. This means that the current continuation parameter has become ill-suited for parameterizing the curve. One way of overcoming this difficulty at turning points is to parameterize the curve by arc length. The augmented Jacobian can be nonsingular throughout the tracing process. However, in practical power system analysis, we always want to get as much useful information as possible during the continuation process. The arc length usually has a geometrical rather than physical meaning, therefore we are often more interested in another important ODE-based predictor, i.e., the tangent parameterized at each step. This is deferred to Section 2.3.4.

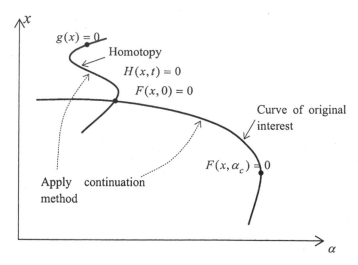

Fig.2.3 Homotopy vs. continuation

Fig.2.3 provides a conceptual view point of homotopy in combination with continuation. Homotopy can be used to get an initial point on the curve of original interest. Continuation method can use this solution to further trace the curve of original interest. As mentioned before homotopy method uses artificial parameter (t in Fig.2.3) to get a solution on the curve of original interest. Continuation method in general uses a natural or physical parameter for the continuation. The efficiency in curve tracing is closely related to the step length control strategy. It is not difficult to choose a workable step size in practice, though some trial and error is often required before the appropriate step size can be found. Step control can often be based on the estimated of the convergence quality of the corrector iteration. In [25] selects step size according to the number of corrector iterations. In general, the step length control scheme is problem dependent.

In practical situations, such as in power systems, saddle node bifurcation, to which out attention will mainly be given, are generic with the collapse type voltage problems. However, in some other situations, other bifurcations might occur more frequently and thus will be of greater interest. For instance, the type of bifurcation that connects equilibria with periodic motion, i.e., Hopf bifurcation, is also generic. Readers interested in problems, such as how to locate Hopf bifurcation point on the traced branch and the related topics are referred to reference [25].

2.3.4 Curve Tracing

Davidenko in his seminal paper [26] proposed that the solving equation Eq.2.12 is equivalent to solving the following differential equation.

$$F_x(x, \lambda) dx / d\lambda = -F_\lambda(x, \lambda) \tag{2.17a}$$

With the initial conditions $x(\lambda) = x_0$, where $F(x_0, \lambda_0) = 0$. One can generate a sequence of solutions for changing λ by numerically solving the differential Eq.2.17b with an appropriate initial value.

$$\frac{dx}{d\lambda} = -[F_x]^{-1} F_\lambda \tag{2.17b}$$

However, here, the singularity of F_x creates numerical problems. Continuation methods can well alleviate this problem.

The continuation algorithm starts from a known solution and uses a predictor–corrector scheme to find subsequent solutions at different λ values.

The Eq.2.17a can be rearranged in the following form

$$F_x(x, \lambda) dx + F_\lambda(x, \lambda) d\lambda = 0 \tag{2.17c}$$

$$\left[F_x(x, \lambda) \quad F_\lambda(x, \lambda)\right] \begin{bmatrix} dx \\ d\lambda \end{bmatrix} = 0$$

Let $T = [\, dx \quad d\lambda \,]^T$, where T is a $n+1$ dimensional vector with $T_{n+1} = \lambda$. T is tangent to the solution branch of Eq.2.17c. Eq.2.17c consists of n equations and $(n+1)$ unknowns. To get a unique solution a normalization of T is needed. For this one can fix one of the elements of T at particular value. For example one can use $e_i^t T = T_i = 1.0$, where e_k is $(n+1)$ dimensional unit vector with k^{th} element equals to unity [25]
The tangent T is the solution of the linear system:

$$\begin{bmatrix} F_x F_\lambda \\ e_k^{tr} \end{bmatrix} z = e_{n+1} \tag{2.18}$$

Provided the full rank condition rank $(F_x, F_\lambda) = n$ holds along the whole branch, the above equation has a unique solution at any point on the branch (k may have to be changed to select a different continuation parameter at a particular step, especially at or near the turning point). Once the tangent vector has been found, the prediction can be easily made. If we define $Y = (x, \lambda)$, then:

$$(\overline{Y}_{j+1}) = (\overline{Y}_j) + \sigma_j T \tag{2.19a}$$

where σ_j designates the step size.

Parameterization and the corrector: Now that a prediction has been made, a method of correcting the approximate solution is needed. Actually the best way to present this corrector is to expand on parameterization, which is vital to the process. Various parameterization techniques are proposed in the mathematical literature. Local parameterization proposed by [27, 28] looks promising and is described here. In local parameterization, the local original set of equations is augmented by one equation that specifies the value of one of the state variables or λ. In local parameterization one can fix $\overline{Y}_k = \eta$ ($1 < k < n+1$). Then we have to solve the following set of equations:

$$\begin{aligned} F(Y) &= 0 \\ \overline{Y}_k - \eta &= 0 \end{aligned} \tag{2.19b}$$

Selection of the continuation parameter corresponds to the variable that has the largest tangent vector component. Therefore $\overline{Y}_k$ at a particular step is the maximum of $(|T_1|, |T_2|, \cdots |T_{n+1}|)$. Now, once a suitable index k and value of η are chosen, a slightly modified Newton-Raphson iterative process can be used to solve the above set of equations. The general form of the iterative corrector process at the jth step is:

$$\begin{bmatrix} F_Y(Y^j) \\ e_k \end{bmatrix} [\Delta Y^j] = \begin{bmatrix} F(Y^j) \\ 0 \end{bmatrix} \tag{2.20}$$

The corrector Jacobian can be seen to have the same form as the predicted Jacobian. Actually the index k used in the corrector is the same as that of used in the predictor and η will be equal to $\overline{Y_k}$, the predicted value Y_k. In the predictor it is made to have a non zero differential change $(d\overline{Y_k} = T_k = \pm 1)$ and in the corrector its value is specified so that the values of other variables can be found.

The step length in Eq.2.19a can be determined by various approaches. The simplest one is by keeping the step length constant. However if we choose very small step length the number of steps needed may be vary large. On the other hand large step lengths may lead to convergence problems. [25] proposed a simple approach for step length selection. Based on this approach the new step length is given by:

$$(\sigma_j)_{new} = (\sigma_j)_{old} N_{opt} / N_j$$

where N_{opt} = optimal number of corrector iterations (this number is 6 for an error tolerance of 10^{-4}) and N_j = Number of iterations needed to approximate the previous continuation step.

With this the $\overline{Y_k}$ value η in Eq.2.19 can be calculated as:

$$\eta = \overline{Y_k^j} + (\sigma_j)_{new}$$

For most of the cases λ is the ideal parameter to choose for tracing. However this parameter creates problems near the fold points. Near the fold point the tangent is normal to the parameter axis. However with local parameterization, near the fold point one can choose the parameter other than λ to avoid these singularity problems. The identification of critical point can be realized by observing the sign change of $d\lambda$.

A one dimensional nonlinear problem is used here to show the basic steps involved in continuation (see Fig.2.4):

Numerical example 2: Consider the following simple example with a single unknown x

$$F(x,\lambda) = x^2 - 3x + \lambda = 0 \qquad (2.21a)$$

The Jacobian is

$$\left[\frac{\partial f}{\partial x} \quad \frac{\partial f}{\partial \lambda}\right] = [(2x-3) \quad 1]$$

Let the base solution (x_0, λ_0) be $(3, 0)$. Then the series of solutions (x_1, λ_1), (x_2, λ_2), can be found using predictor-corrector continuation as below:

Continuation step 1:

Predictor

To start with, let λ be the continuation parameter. Calculate the tangent vector as below: (here the index k is equal to 2).

$$\begin{bmatrix} (2x_0 - 3) & 1 \\ 0 & 1 \end{bmatrix} \begin{bmatrix} dx \\ d\lambda \end{bmatrix} = \begin{bmatrix} 0 \\ 1 \end{bmatrix}$$

$$\Rightarrow \begin{bmatrix} 3 & 1 \\ 0 & 1 \end{bmatrix} \begin{bmatrix} dx \\ d\lambda \end{bmatrix} = \begin{bmatrix} 0 \\ 1 \end{bmatrix}$$

$$\Rightarrow dx = -\frac{1}{3} \quad \text{and} \quad d\lambda = 1$$

Predict the next solution by solving:

$$\begin{bmatrix} \overline{x}_1 \\ \overline{\lambda}_1 \end{bmatrix} = \begin{bmatrix} x_0 \\ \lambda_0 \end{bmatrix} + \sigma \begin{bmatrix} dx \\ d\lambda \end{bmatrix}$$

where σ is a scalar designating step size (say 0.5). Thus the predicted solution $(\overline{x}_1, \overline{\lambda}_1)$ becomes $(2.8333, 0.5)$.

Continuation step 2:
Corrector

Correct the predicted solution by solving:

$$-\begin{bmatrix} (2\overline{x}_1 - 3) & 1 \\ 0 & 1 \end{bmatrix} \begin{bmatrix} \Delta x \\ \Delta \lambda \end{bmatrix} = \begin{bmatrix} f(\overline{x}_1, \overline{\lambda}_1) \\ 0 \end{bmatrix}$$

$$\Rightarrow -\begin{bmatrix} 2.6666 & 1 \\ 0 & 1 \end{bmatrix} \begin{bmatrix} \Delta x \\ \Delta \lambda \end{bmatrix} = \begin{bmatrix} 0.0277768 \\ 0 \end{bmatrix}$$

$$\Rightarrow \Delta x = -0.0104165 \quad \text{and} \quad \Delta \lambda = 0$$

Repeat these correction iterations until reasonable accuracy is obtained (say $\varepsilon = 0.0001$). Now, $\max\{\Delta x, \Delta \lambda\} > \varepsilon$. So update x_1 and λ_1 and repeat the corrector iteration.

$$\Rightarrow \overline{x}_{1,new} = \overline{x}_1 + \Delta x = 2.8229164$$

$$\overline{\lambda}_{1,new} = \overline{\lambda}_1 + \Delta \lambda = 0.5$$

Continuation step 3:
Corrector iteration:

$$-\begin{bmatrix} (2\overline{x}_{1,new} - 3) & 1 \\ 0 & 1 \end{bmatrix} \begin{bmatrix} \Delta x \\ \Delta \lambda \end{bmatrix} = \begin{bmatrix} F(\overline{x}_{1,new}, \overline{\lambda}_{1,new}) \\ 0 \end{bmatrix}$$

$$\Rightarrow \Delta x = -0.00004075 \text{ and } \Delta \lambda = 0$$

Now, $\max\{\Delta x, \Delta \lambda\} < \varepsilon$. So stop the corrector iterations. After the first continuation step, the point (x_1, λ_1) is equal to $(2.8228757, 0.5)$. Repeat the entire process until we reach the critical point. For this example, the critical point it $(1.5, 2.25)$.

λ versus x curve for the example is shown in the Fig.2.4. For the predicted solutions at points (1) and (2), we can choose λ as the continuation parameter (i.e., fix λ at that particular value) and converge on to the curve with corrector iterations. But at (3), λ can not be a continuation parameter, as there is no solution for that value of λ. At this point (i.e. when we are close to the critical point), we use local parameterization technique and choose x as the continuation parameter and solve for the system. This can be clearly observed in the example we considered. We know the solution at the fold point as $(1.5, 2.25)$. Consider the augmented Jacobian of the continuation process

$$J_{aug} = \begin{bmatrix} (2x - 3) & 1 \\ 0 & 1 \end{bmatrix}$$

At $x = 1.5$, $det\{J_{aug}\} = 0$. So the method diverges near the critical point, if λ is the continuation parameter. If we fix the value of x, instead of λ, then

$$J_{aug} = \begin{bmatrix} (2x - 3) & 1 \\ 1 & 0 \end{bmatrix}$$

$det\{J_{aug}\} \neq 0$. Then we can solve for the system.

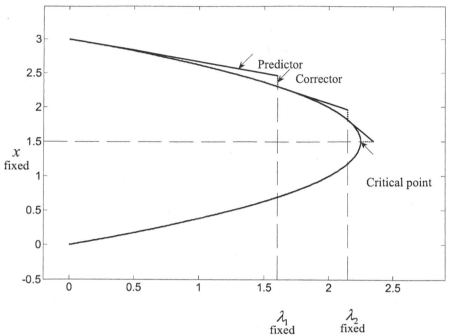

Fig.2.4 Illustration of predictor-corrector scheme

2.3.5 Direct method in computing the Saddle node bifurcation point: a one step continuation

In Section 2.3.4, discussion has been given to show that the tracing of a curve can be done via continuation. We've noticed that, on the traced curve, a particular point, namely, the critical point, or sometimes called the fold point (also related to saddle node bifurcation), is often of greater interest.   If we are only interested in locating this point with respect to λ_c, or say, we are interested in the maxim allowable variation of λ where the corresponding linearization (Jacobian) is singular, we have yet another approach available, i.e., the direct method.

When the Jacobian becomes singular, $F(x, \lambda) = 0$ can not be solved by regular Newton–Raphson method in the present form. To avoid this singularity several methods have been published in the mathematical literature [29, 30, 31]. In these references the authors cleverly augmented the original system of equations in such a way that that for this enlarged system, the fold point becomes regular. If the fold point is mathematically characterized by the steady state Jacobian F_x having a simple and unique zero eigenvalue, with nonzero right eigenvector h and left eigenvector w, then

$$G(Y) = \begin{bmatrix} F(x,\lambda) \\ h_k - 1 \\ F_x(x,\lambda)h \end{bmatrix} = 0 \qquad\qquad (2.21b)$$

where (x_c, λ_c) is a fold point of $F(x_c, \lambda) = 0$. This procedure basically augments the original equations of $F(x, \lambda) = 0$ by $F_x(x, \lambda)h = 0$, with $h_k = 1$. This augmentation makes the Jacobian G_x of enlarged system $G(Y)$ non-singular and guarantees a solution. The proof can be found in [32]. This approach has some drawbacks. The dimension of the nonlinear set of equations to be solved is twice that of the original number. The approach requires a good estimate for the vector h. However, convergence of the direct method is very fast if the initial operating point is close to the turning point. The enlarged system can be solved in such a way that it requires the solution of $n \times n$ (n is the dimension of the Jacobian $F_x(x, \lambda)$) linear systems, each with the same matrix. This method needs only one LU decomposition. At this turning point, rank $F_x(x, \lambda) = n-1$ and $F_\lambda (x_c, \lambda_c)$ ε *range* $F_x (x_c, \lambda_c)$, that is rank $F_x (x_c, \lambda_c)/ F_\lambda (x_c, \lambda_c) = n$. These are called transversality conditions. Depending on the type of transversality condition, different types of static bifurcations can occur. Fold or saddle node is generic or the most commonly occurring static bifurcation. Table 2.1 summarizes the type of static bifurcation and corresponding transversality condition for a one-dimensional scalar system. Details can be found in Wiggins book [33]. The application of this method to power system voltage stability is reported by [34, 35].

Table 2.1 Static bifurcation types

Bifurcation Type	Transversality Condition	Prototype Equation	Bifurcation Diagram
Fold	$\dfrac{\partial F}{\partial \lambda} \neq 0; \dfrac{\partial^2 F}{\partial x^2} \neq 0$	$\lambda - x^2 = 0$	
Transcritical	$\dfrac{\partial^2 F}{\partial \lambda \partial x} \neq 0; \dfrac{\partial^2 F}{\partial x^2} \neq 0$	$\lambda x - x^2 = 0$	

| Pitch Fork | $\dfrac{\partial^2 F}{\partial\lambda\partial x} \neq 0; \dfrac{\partial^2 F}{\partial x^2} \neq 0$ | $\lambda x - x^3 = 0$ | |

– – – Unstable mode ——— Stable mode

Numerical example 3: In Eq.2.21b, the original system of equations is augmented in such a way that for the enlarged system, the turning point becomes regular. Solving for Eq.2.21b will yield the desired fold point. Related to numerical example 2 given above, the enlarged system of equations is:

$$
\begin{cases} x^2 - 3x + \lambda = 0 \\ 2x - 3 = 0 \\ v = 1 \end{cases} \Rightarrow \begin{cases} x^2 - 3x + \lambda = 0 \\ 2x - 3 = 0 \end{cases} \tag{2.22}
$$

Solving the above two equations, we directly get the critical point $(x_c, \lambda_c)^{tr}$ $= (1.5, 2.25)^{tr}$.

Advantages:

The direct method can find the critical point where the Jacobian is singular by solving the enlarged system of power flow equations in one step.   The left and right eigenvectors produced in the direct approach carry very important information. For instance, it was shown that, at saddle node bifurcations, the right eigenvector corresponding to the zero eigenvalue gives the trajectory of the system state variables [16]. The left eigenvector can be used to construct a normal vector [17, 36, 37, 38] at the bifurcations hypersurface.

Limitations:

In the direct approach, for a successful convergence, a good initial guess is needed. This method basically doubles the number of equations to be solved.   However, some of these shortcomings can be overcome by following the approach proposed in reference [30]. In that paper, the authors explored the structure of equation Eq.2.21b. It's shown that the whole system can be resolved into four linear subsystems with the same coefficient matrix. Reference [34] applied this method to power system voltage stability studies.

When one is beset with the lack of good starting points for the Newton type iterative methods in solving nonlinear equations, or when one needs to lead a parameter study of nonlinear system equilibrium problems, it would probably be advisable to turn homotopy and continuation methods. Examples shown in this section manifests the applicability of the technique to engineering problems. This review does not present and exhaustive survey but a compact text on continuation methods. Readers interested in continuation, bifurcation, and related numerical methods may find the following references [25, 27, 39, 40] very helpful.

2.4 Hopf Bifurcation

2.4.1 Existence of Hopf bifurcation point

If (i) $F(x_c, \lambda_c) = 0$, (ii) the Jacobian matrix $(\partial F/\partial x)$ has a simple pair of purely imaginary eigenvalues, $\mu(\lambda_c) = \pm jw$, (iii) $d(\text{Re}(\mu(\lambda_c)))/d\lambda \neq 0$.
(Marsden & McCracken [41], Hassard *et al* [42].)

Then there is a birth or death of limit cycles at (x_c, λ_c) depending on the sign of derivative in (iii). λ_c is the value of the parameter at which Hopf bifurcation occurs. Requirement (iii) guarantees there is a transversal crossing of the imaginary axis by the pair of complex conjugate eigenvalues. Numerical determination of the Hopf bifurcation point involves estimation of the point (x_c, λ_c). A costly way of identifying the point is to evaluate all the eigenvalues of the Jacobian matrix. However, as in the static approach there are efficient ways of identifying the Hopf point by direct methods as well as by indirect methods.

2.4.1.1 Direct methods

Direct methods [40] calculate the Hopf point by solving one single suitably chosen equation. At the Hopf point, one pair of complex eigenvalues crosses the imaginary axis. Let this pair be:

$$\mu(\lambda) = \alpha(\lambda) - j\beta(\lambda)$$

with

$$\alpha(\lambda_c) = 0; \ \beta(\lambda_c) = 0; \ d\alpha(\lambda_c)/d\lambda = 0$$

For an eigenvalue μ of the Jacobian matrix $F_x [= \partial F / \partial x]$, the following equation is valid

$$F_x W = \mu W \qquad (2.23)$$

where $W = u + jv$ is an eigenvector corresponding to the eigenvalue μ. Since $\alpha(X_c) = 0$, Eq.2.23 can be written as

$$F_x (u + jv) = (+j\beta)(u + jv)$$

$$F_x u + jF_x v = -\beta v + j\beta u$$

$$F_x v + \beta u = 0 \qquad (2.24)$$

$$F_x v - \beta u = 0 \qquad (2.25)$$

where u and v are vectors of dimension n. We have in fact $3n$ nonlinear algebraic Eqs.2.24 and 2.25 and $F(x,\lambda)=0$ with $3n+2$ unknowns ($x_1, x_2, \cdots,$ $x_n, u_1, u_2, \cdots, u_n, v_1, v_2, \cdots, v_n, \lambda, \beta$). However the other two unknowns can be obtained by putting two normalizing conditions that force W to be non-zero. This means that practically we can choose two components of the vectors u and v arbitrarily. The Newton iterations method can be effectively used to solve this $3n$ by the $3n$ system to get the Hopf point. An efficient algorithm based on the direct approach is provided by [43]. The application of the boundary value problem for direct computation of the Hopf points was proposed by Seydel [25].

2.4.1.2 Indirect methods

The Hopf bifurcation point (x_c, λ_c) can also be located by an indirect approach. This can be achieved by obtaining the information collected during any continuation method described before, i.e., an iteration technique is used to solve the algebraic equation $\mathrm{Re}(\mu(\lambda)) = 0$ by means of the secant method. A change of sign of the real part $\alpha(\lambda)$ indicates that λ_c has been passed. Therefore the check $\alpha(\lambda j) \, \alpha(\lambda j - 1) < 0$ should be performed after each continuation step $\lambda_{j-1} \to \lambda_j$.

A good comparison of various methods of computing Hopf bifurcating points is given by [44]. Application of Hopf bifurcation to power system problems can be found in [12, 14, 22, 45].

2.5 Complex Bifurcation

Further variation of the parameter beyond the Hopf point may lead to other complex phenomena; basically one has to trace the monodromy matrix of a periodic orbit for different values of the parameter. The stability of periodic solution is determined by Floquet multipliers which are the eigenvalues of the monodromy matrix. For a particular value of λ, the monodromy matrix has n-Floquet multipliers. The magnitude of one of them is always equal to unity. The other n-1 Floquet multipliers determine (local) stability by the following rule [25, 46].

- $x(t)$ is stable if $| \mu_j | < 1$, for $j = 1, ..., n$-1;
- $x(t)$ is unstable if $| \mu_j | > 1$, for some j.

On the stable periodic orbit, the n-1 multipliers are always inside the unit circle. The multipliers are the functions of the parameter under consideration. When we vary the parameter, some of the multipliers may cross the unit circle. The multiplier crossing the unit circle is called the critical multiplier. Different types of branching occur depending on where a critical multiplier or pair of complex conjugate multipliers leaves the unit circle. Three associated types of branching are (i) the critical multiplier goes outside the unit circle along the positive real axis, with $| \mu(p_c) | = 1$, (ii) the multiplier goes outside the unit circle along the negative real axis with $| \mu(p_c) | = -1$ and (iii) a pair of complex conjugate multipliers crosses the unit circle with a non-zero imaginary part. All these types refer to a loss of stability when λ passes through λ_c. (On the other hand, if a critical multiplier enters the unit circle, the system gains stability.) In the case (i) typically, turning points of the periodic orbit occur with a gain or loss of stability. Transcritical or pitchfork type bifurcations in periodic orbits are also possible for this case. In the case (ii), the system oscillates with period two. In the case (iii), the phenomenon of bifurcation into a torus occurs, which is also called secondary Hopf bifurcation, or generalized Hopf bifurcation. The period doubling bifurcation often occurs repeatedly which generally leads to chaos. Lyapunov exponents are generally used to identify the chaos [47]. The exponential serves as a measure for exponential divergence or contraction of nearby trajectories. Chaos is characterized by at least one positive Lyapunov exponent, which reflects a stretching into one or more directions. In general, chaos has the following ingredients [47]: (i) the underlying dynamics is deterministic, (ii) no external noise has been introduced, (iii) seemingly erratic behavior of individual trajectories depends sensitively on small changes of initial conditions; (iv) in contrast to a single trajectory, some global characteristics are obtained by averaging over many trajectories or over a long time (e.g., a positive Lyapunov exponent) that do not depend on initial conditions; (v) when a parameter is tuned, the erratic state is reached via a sequence of events, including the appearance of one or more sub-harmonics. In the last few years, a great number of conferences and workshops devoted to chaotic dynamics have been organized. In most of them, papers by researchers from various branches of science and engineering have been presented. Research in chaos is well documented by [47]. Numerical methods to identify chaos can be found in [48]. Observations of chaos in power systems are reported in [12, 13, 49]. Fig2.5 gives the overall possible bifurcation scenario.

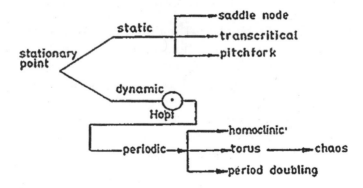

Fig. 2.5 List of possible bifurcations

References

[1]   Blaquiere, A., Nonlinear system analysis New York: Academic Press 1966

[2]   Jordan, D. W., Smith, P., Nonlinear ordinary differential equations. Oxford: University Press 1977

[3]   Reyn, J.W., Classification and description of the singular points of a system of three linear differential equations, J. Appl. Math. Phys. 15: 540-555, 1964

[4]   Ajjarapu, V., "The Role of Bifurcation and Continuation Methods in the Analysis of Voltage Collapse," Sadhana , Vol. 18, No. 5, pp. 829-841, September 1993

[5]   Guckenheimer, J., Holmes, P. J., Nonlinear oscillations, dynamical systems, and bifurcation of vector fields. New York: Springer-Verlag 1983

[6]   Carr, J., Applications of center manifold theory. Applied Math. Sciences. Vol. 35 New York: Springer-Verlag 1981

[7]   Haken, H., Synergetics, an introduction 3rd ed. Berlin: Springer 1983

[8]   Arnold, V. I., Lectures on bifurcation in versal families. Russ. Math. Survey 27:54-123 1972

[9]   Bruno, A. D., Local methods in nonlinear differential equations Berlin: Springer-Verlag 1989

[10]   Thapar, J., Vittal, V., Kliemann, W., Fouad, A.A., Application of the normal form of vector fields to predict interarea separation in power systems. IEEE Transactions on Power Systems, Vol.12: 844 - 850 May 1997

[11]   Dobson, I., Barocio, E., Scaling of normal form analysis coefficients under coordinate change. IEEE Transactions on Power Systems Vol. 19: 1438 - 1444 Aug. 2004

[12]   Ajjarapu, V., Lee B Bifurcation theory and its application to nonlinear dynamical phenomenon in an electric power system. IEEE Trans. Power System 7:424-431 1992

[13]   Chiang, H. D., Liu, C. W., Varaiya, P. P., Wu,F.F., Lauvy,M.G., Chaos in a simple power system, IEEE Transactions on Power Systems, Vol. 8, no.4, pp.1407-1417, Nov.1993

[14]   Rajgopalan, C., Sauer, P. W., Pai, M. A., Analysis of voltage control system exhibiting Hopf bifurcation. IEEE proceedings of the 28th conference on Decisions and Control New York 1989

[15]   Riks, E., An incremental approach to the solution of snapping and buckling problems.   Int. J. Solids Struct. 15: 529-551 1979

[16] Dobson, I., Chiang. H. D., Toward a theory of voltage collapse in electric power systems. Systems and control letters, Vol. 13: 253-262 1990

[17] Dobson, I., Observations on the geometry of saddle node bifurcations and voltage collapse in electric power systems. IEEE Trans. on CAS, part I, Vol. 39 240-243 1992

[18] Halvacek, V(ed.)., Chaos (Singapore: World Scientific) 1990

[19] Moon, F. C., Chaotic Vibration (New York: Wiley) 1987

[20] Thomson, J. M. T., Stewart, H. B., Nonlinear dynamical systems and chaos (New York: Springer Verlag) 1986

[21] Holmes,.P. J(ed.)., New approaches to nonlinear problems in dynamics Philadelphia: Siam 1980

[22] Venkatasubramanian, V., Juan, Li., Study of Hopf bifurcations in a simple power system model. Proceedings of the 39th IEEE Conference on Decision and Control, Vol. 4 Page:3075 - 3079 Dec. 2000

[23] Allgower, E. L., Georg, k., Numerical continuation methods Berlin: Springer 1990

[24] Kincaid, D., Cheney, W., Numerical anlaysis. California: Pacific grove 1991

[25] Seydel, R., From equilibrium to chaos (New York: Elsevier Science) 1988

[26] Davidenko, D., On a new method of numerically integrating a system of nonlinear equations. Dokl., Akad. Nauk, USSR 88:601-604 1953

[27] Rheinboldt, W. C., Numerical analysis of parametrized nonlinear equations.   (New York: John Wiley & Sons) 1986

[28] Rheinboldt, W. C., Burkhardt, J. V., A locally parameterized continuation process.   ACM Trans. Math. Software 9: 215-235 1983

[29] Abbot, J. P., An efficient algorithm for the determination of certain bifurcation points. J. Comput. Appl. Math. 4: 19-27 1978

[30] Moore, G., Spence, A., The calculation of turning points of nonlinear equations.   SIAM J. Numer. Anal. 17: 567-576 1980

[31] Griewank, A., Reddien, G. W., Characterization and computation of generalized turning points. SIAM J. Numer. Anal. 21: 176-185 1984

[32] Seydel, R., Numerical computation of periodic orbits that bifurcate from the stationary solution of ordinary differential equation. Appl. Math. Comput. 9: 257-271 1979

[33] Wiggins, S., Introduction to Applied Nonlinear Dynamical Systems and Chaos New York: Springer-Verlag 1990

[34] Ajjarapu, V., Identification of steady state voltage stability in power systems. Int. J. Energy Syst. 11: 43-46 1991

[35] Alvarado, F. L., Jung, T. H., Direct detection of voltage collapse conditions. Proceedings: Bulk Power System Voltage Phenomenon – Voltage Stability and Security, EPRI EL-6183, Project 2473-21, Electric Power Research Institute 1989

[36] Dobson, I., Lu, L., Computing an optimum direction in control space to avoid saddle node bifurcations and voltage collapse in electric power systems. IEEE Trans. on automatic control, Vol. 37 1616-1620 1992

[37] Dobson, I., Computing a closest bifurcation instability in multidimensional parameter space. Journal of nonlinear science, Vol. 3, No. 3: 307-327 1993

[38] Dobson, I., Lu, L., New methods for computing a closest saddle node bifurcation and worst case load power margin for voltage collapse. IEEE Trans. on power systems, Vol. 8: 905-913 1993

[39] Allgower, E. L., Georg, k., Numerical continuation methods Berlin: Springer 1990

[40] Kubicek, M., Marek, M., Computational methods in bifurcation theory and dissipative structures (New York: Springer-Verlag) 1983

[41] Marsden, J. E., McCracken, M., The hopf bifurcation and its application (New York: Springer) 1976

[42] Hassard, B. D., Kazirinoff, N. D., Wan, Y. H., Theory and applications of Hopf bifurcation Cambridge: University Press 1981

[43] Griewank, A., Reddien, G., The calculation of Hopf bifurcation by direct method. Inst. Math. Appl. J. Numer. Anal. 3: 295-303 1983

[44] Roose, D., An algorithm for the computation of Hopf bifurcation points in comparison with other methods. J. Comput. Appl. Math. 12 & 13: 517-529 1985

[45] Rajagopalan, C., Lesieutre, B., Sauer, P. W., Pai, M.A., Dynamic aspects of voltage/power characteristics, IEEE Trans. on power systems, Vol. 7, no.3, pp. 990-1000, Aug.1992

[46] Arnold, V. I., Geometric methods in the theory of ordinary differential equations New York: Springer 1983

[47] Lin, H. B., Chaos II World Scientific Pub. Co. 1990

[48] Parker, T. S., Chua, L., Practical numerical algorithms for chaotic systems. (New York: Springer) 1989

[49] Nayfeh, M. A., Hamdan, A. M. A., Nayfeh, A. H., Chaos and instability in a power system: primary resonant case. Nonlinear Dynamics 1: 313-339 1990

[50] Pai, M.A., Sen Gupta D.P, and Padiyar K.R. " Small Signal Analysis of Power Systems " Originally Published by Narosa Publishing House and Co-Published by Alpha Science International Limited, Oxford, U.K., 2004

3 Continuation Power Flow

3.1 Introduction

As mentioned in the previous chapter, the continuation method is a mathematical path-following methodology used to solve systems of nonlinear equations. The numerical derivation of this method is shown in [1]. Using the continuation method, we can track a solution branch around the turning point without difficulty. This makes the continuation method quite attractive in approximations of the critical point in a power system. The continuation power flow captures this path-following feature by means of a predictor-corrector scheme that adopts locally parameterized continuation techniques to trace the power flow solution paths. The next sections explain the principles of continuation power flow.

3.2 Locally Parameterized Continuation

A parameterization is a mathematical means of identifying each solution on the branch, a kind of measure along the branch. When we say "branch," we refer to a curve consisting of points joined together in $n+1$ dimensional space that are solutions of the nonlinear equations

$$F(x,\lambda) = 0 \tag{3.1}$$

This equation is obtained by introducing a load parameter, λ, into the original system of nonlinear equations, $F(x) = 0$. For a range of values of λ, it is quite possible to identify each solution on the branch in a mathematical way [2]. But not every branch can be parameterized by an arbitrary parameter. The solution of Eq.3.1 along a given path can be found for each value of λ, although problems arise when a solution does not exist for some maximum possible λ value. At this point, one of the state variables, x_i, can be used effectively as the parameter to be varied, choice of which is determined locally at each continuation step. Thus, the method is designated as the locally parameterized continuation. In summary, local parameterization allows not only the added load parameter λ, but also the state variables to be used as continuation parameters.

3.3 Formulation of Power Flow Equations

To apply locally parameterized continuation techniques to the power flow problem, the power flow equations must be reformulated to include a load parameter, λ. This reformulation can be accomplished by expressing the load and the generation at a bus as a function of the load parameter, λ. Thus, the general forms of the new equations for each bus i are

$$\Delta P_i = P_{Gi}(\lambda) - P_{Li}(\lambda) - P_{Ti} = 0 \tag{3.2}$$

$$\Delta Q_i = Q_{Gi} - Q_{Li}(\lambda) - Q_{Ti} = 0 \tag{3.3}$$

where

$$P_{Ti} = \sum_{j=1}^{n} V_i V_j y_{ij} \cos(\delta_i - \delta_j - \gamma_{ij})$$

$$Q_{Ti} = \sum_{j=1}^{n} V_i V_j y_{ij} \sin(\delta_i - \delta_j - \gamma_{ij})$$

and $0 \le \lambda \le \lambda_{critical}$. $\lambda = 0$ corresponds to the base case, and $\lambda = \lambda_{critical}$ to the critical case. The subscripts L, G and T respectively denote bus load, generation, and injection. The voltage at bus i is $V_i \angle \delta_i$, and $y_{ij} \angle \gamma_{ij}$ is the $(i, j)^{th}$ element of the system admittance matrix $[y_{BUS}]$.

To simulate different load change scenarios, the P_{Li} and Q_{Li} can be modified as

$$P_{Li}(\lambda) = P_{Li0} + \lambda[K_{Li} S_{\Delta BASE} \cos(\psi_i)] \tag{3.4}$$

$$Q_{Li}(\lambda) = Q_{Li0} + \lambda[K_{Li} S_{\Delta BASE} \sin(\psi_i)] \tag{3.5}$$

where $S_{\Delta BASE} \cos(\Psi_i) = P_{Li0}$ and $Q_{Li0} = S_{\Delta BASE} \sin(\Psi_i)$.

Let $Q_{Li0} = P_{Li0} \tan(\Psi_i)$, then Eqs.3.4 and 3.5 can be rewritten as

$$P_{Li}(\lambda) = P_{Li0}[1 + \lambda K_{Li}]$$

$$Q_{Li}(\lambda) = P_{Li0} \tan(\Psi i)[1 + \lambda K_{Li}]$$

where

- P_{Li0}, Q_{Li0} = original load at bus i, active and reactive respectively;

- K_{Li} = multiplier designating the rate of load change at bus i as λ changes;

- ψ_i = power factor angle of load change at bus i;

- $S_{\Delta BASE}$ = apparent power, which is chosen to provide appropriate scaling of λ.

The active power generation can be modified to

$$P_{Gi}(\lambda) = P_{Gi0}(1 + \lambda K_{Gi}) \tag{3.6}$$

where

- P_{Gi0} = active generation at bus i in the base case;

- K_{Gi} = constant specifying the rate of change in generation as λ varies.

Now if F is used to denote the entire set of equations, then the problem can be expressed as a set of nonlinear algebraic equation represented by Eq.3.1, with $x = [\delta, V]^T$. The predictor corrector continuation process can then be applied to these equations.

3.4 The Predictor-corrector Process

The first task in the predictor process is to calculate the tangent vector. This can be obtained from

$$[\underline{F}_\delta, \underline{F}_v, \underline{F}_\lambda] \begin{bmatrix} d\underline{\delta} \\ d\underline{V} \\ d\lambda \end{bmatrix} = 0$$

On the left side of the equation is a matrix of partial derivatives multiplied by the vectors of differentials. The former is the conventional power flow Jacobian augmented by one column (F_λ), whereas the latter $T = [d\underline{\delta}, d\underline{V}, d\lambda]^T$ is the tangent vector being sought. Normalization must be imposed to give $\underline{t}$ a nonzero length. One can use, for example

$$e_k^T \underline{t} = t_k = 1$$

where e_k is an appropriately dimensioned row vector with all elements equal to zero except the kth, which is equal to one. If the index k is chosen properly, letting $t_k = \pm 1.0$ imposes a nonzero norm on the tangent vector and guarantees that the augmented Jacobian will be nonsingular at the point of maximum possible system load [3]. Thus, the tangent vector is determined as the solution of the linear system

$$\begin{bmatrix} \underline{F}_\delta & \underline{F}_v & \underline{F}_\lambda \\ & e_k & \end{bmatrix} [\underline{t}] = \begin{bmatrix} 0 \\ \pm 1 \end{bmatrix} \tag{3.7}$$

Once the tangent vector has been found by solving Eq.3.7, the prediction can be made as

$$\begin{bmatrix} \underline{\delta}^* \\ \underline{V}^* \\ \lambda^* \end{bmatrix} = \begin{bmatrix} \underline{\delta} \\ \underline{V} \\ \lambda \end{bmatrix} + \sigma \begin{bmatrix} d\underline{\delta} \\ d\underline{V} \\ d\lambda \end{bmatrix}$$

where '*' denotes the predicted solution, and σ is a scalar designating step size.

After the prediction is made, the next step is to correct the predicted solution. As mentioned in Chapter 2, the technique used here is local parameterization, whereby the original set of equations is augmented by one equation specifying the value of one of the state variables. In equation form, this relation is expressed as

$$\begin{bmatrix} F(x) \\ x_k - \eta \end{bmatrix} = 0, \qquad x = \begin{bmatrix} \delta \\ V \\ \lambda \end{bmatrix}$$

where η is an appropriate value for the kth element of x. Once a suitable index k and the value of η are specified, a slightly modified N-R power flow method (altered only by one additions equation and one additional state variable) can be used to solve the set of equations. This procedure provides the corrector needed to modify the predicted solution found in the previous section.

3.4.1 Selecting the continuation parameter

The best method of selecting the correct continuation parameter at each step is to select the state variable (change the underlined to variable) with the largest tangent vector component. In short, we select the state variable (change the underlined to variable) evidencing the maximum rate of change near a given solution. To begin with, λ is a good choice, and subsequent continuation parameters can be evaluated as:

$$x_k : |t_k| = \max\{|t_1|, |t_2|, \cdots, |t_m|\} \tag{3.8}$$

Here, t is the tangent vector. After the continuation parameter is selected, the proper value of either +1 or -1 should be assigned to t_k in the tangent vector calculation.

3.4.2 Identifying the critical point

To find the stopping criterion for the continuation power flow, we must determine whether the critical point has been reached. This can be done easily because the critical point is the point at which maximum loading (and hence maximum λ) occurs before decreasing. For this reason, at the critical point, the tangent vector component corresponding to λ (which is $d\lambda$) is zero and becomes negative once it passes the critical point. Thus, the sign of the $d\lambda$ component tells us whether the critical point has been passed or not.

The previous paragraphs summarize the basic continuation power flow. More details can be found in [4].

3.5 Examples

__Two bus example__: constant power load: the above approach is first demonstrated through a simple two bus example as shown in Fig.3.1.

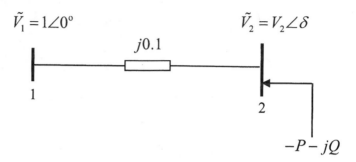

Fig.3.1 Two bus system

For this two bus example, the power flow equation at bus 2 can be formulated as:

$$(-P - jQ)^* = V_2 \angle -\delta(Y_{21}V_1\angle\theta_{21} + Y_{22}V_2\angle\delta)$$

Suppose

- The voltage at the generator bus is: $\tilde{V}_1 = 1\angle 0$

- The voltage at the load bus is: $\tilde{V}_2 = V_2\angle\delta$

- The load is: $P + jQ$, so the injected power is: $-P - jQ$

- The load power factor keeps constant.

By introducing parameters λ and K, we can represent load increase scenario at bus 2 as follows:

$$P = P_0 * (1 + \lambda K)$$
$$Q = Q_0 * (1 + \lambda K)$$

where $Q_0 = P_0 * \tan(\psi)$ and is the constant load changing factors specified for bus 2.

Then we get two equations corresponding to real power and reactive power:

$$0 = P_0 * (1 + \lambda K) + Y_{21} V_2 \cos(\theta_{21} - \delta) + Y_{22} V_2^2 \cos(\theta_{22}) = f_1(\delta, V_2, \lambda) \quad (3.9)$$

$$0 = Q_0 * (1 + \lambda K) - Y_{21} V_2 \sin(\theta_{21} - \delta) - Y_{22} V_2^2 \sin(\theta_{22}) = f_2(\delta, V_2, \lambda) \quad (3.10)$$

Now the original Jacobin matrix can be expressed as follows:

$$J_0 = \begin{bmatrix} \dfrac{\partial f_1}{\partial \delta} & \dfrac{\partial f_1}{\partial V_2} \\[2ex] \dfrac{\partial f_2}{\partial \delta} & \dfrac{\partial f_2}{\partial V_2} \end{bmatrix} = \begin{bmatrix} Y_{21} V_2 \sin(\theta_{21} - \delta) & Y_{21} \cos(\theta_{21} - \delta) + 2 Y_{22} V_2 \cos(\theta_{22}) \\ Y_{21} V_2 \cos(\theta_{21} - \delta) & -Y_{21} \sin(\theta_{21} - \delta) - 2 Y_{22} V_2 \sin(\theta_{22}) \end{bmatrix}$$

In the example, the system parameters are given as follows: $K = 1.0$, $P_0 = 0.1$, $\cos(\psi) = 1.0$, $Y_{12} = Y_{22} = 10$, $\theta_{12} = 90°$, $\theta_{22} = -90°$.

Suppose we start from the following initial point:

$$V_2 = 1.004, \quad \delta = 0.075°, \quad \lambda = 0$$

Using the initial guess of V_2, a prediction of the next solution can be made by taking an appropriately sized step in a tangent direction to the solution path. The tangent vector can be calculated using the augmented Jacobin matrix:

$$J_{Aug} = \begin{bmatrix} J_0 & \dfrac{\partial f}{\partial \lambda} \\[2ex] e_k \end{bmatrix} = \begin{bmatrix} \dfrac{\partial f_1}{\partial \delta} & \dfrac{\partial f_1}{\partial V_2} & P_0 * K \\[2ex] \dfrac{\partial f_2}{\partial \delta} & \dfrac{\partial f_2}{\partial V_2} & Q_0 * K \\[2ex] & e_k & \end{bmatrix} \quad (3.11)$$

where e_k is an appropriately dimensioned row rector with all elements equal to zero except the kth, which equals one. If the index k is chosen correctly, the augmented matrix is nonsingular. In the beginning, $k = 3$ is chosen which corresponds to the parameter λ. So J_{Aug} is

$$
J_{Aug} = \begin{bmatrix} 10.0392 & 0.13 & 0.1 \\ 0.1305 & 10.0808 & 0 \\ 0 & 0 & 1 \end{bmatrix}
$$

Define the tangent vector as:

$$
t = \begin{bmatrix} d\delta & dV_2 & d\lambda \end{bmatrix}^T
$$

During the prediction process, we have:

$$
J_{Aug} * t = \begin{bmatrix} 0 & 0 & 1 \end{bmatrix}^T \tag{3.12}
$$

In the beginning the tangent vector is

$$
t = \begin{bmatrix} -0.01 & 0.0001 & 1 \end{bmatrix}^T
$$

With this tangent vector, we get the predicted solution:

$$
\begin{bmatrix} \delta^{k+1} \\ V_2^{k+1} \\ \lambda^{k+1} \end{bmatrix} = \begin{bmatrix} \delta^k \\ V_2^k \\ \lambda^k \end{bmatrix} + \sigma \begin{bmatrix} d\delta^k \\ dV_2^k \\ d\lambda^k \end{bmatrix}
$$

where σ is a specified step length (we start at an initial step length σ of 0.3. Subsequent step lengths can be determined according to the procedure described in Chapter 2). So we have

$$
\begin{bmatrix} \overline{\delta}^1 \\ \overline{V}_2^1 \\ \overline{\lambda}^1 \end{bmatrix} = \begin{bmatrix} \delta^0 \\ V_2^0 \\ \lambda^0 \end{bmatrix} + \sigma \begin{bmatrix} d\delta^0 \\ dV_2^0 \\ d\lambda^0 \end{bmatrix} = \begin{bmatrix} 0.013 \\ 1.004 \\ 0 \end{bmatrix} + 0.3 * \begin{bmatrix} -0.01 \\ -0.0001 \\ 1 \end{bmatrix} = \begin{bmatrix} 0.01 \\ 1.004 \\ 0.3 \end{bmatrix}
$$

Now that a prediction has been achieved, we can use this predicted solution as an initial guess for the corrector. We use the local parameterization method. Substituting these values into Eqs.3.9 and 3.10, we get the mismatch: $\Delta f_1^k, \Delta f_2^k$. Here we let λ^k be constant and apply the same augmented Jacobin matrix, to obtain the corrector:

$$
\begin{bmatrix} \Delta\delta^{k+1} \\ \Delta V_2^{k+1} \\ \Delta\lambda^{k+1} \end{bmatrix} = -J_{Aug}^{-1} * \begin{bmatrix} \Delta f_1^k \\ \Delta f_2^k \\ 0 \end{bmatrix}
$$

So the corrected solution is:

$$
\begin{bmatrix} \delta^{k+1} \\ V_2^{k+1} \\ \lambda^{k+1} \end{bmatrix} = \begin{bmatrix} \delta^{k+1} \\ V_2^{k+1} \\ \lambda^{k+1} \end{bmatrix} + \sigma \begin{bmatrix} \Delta\delta^{k+1} \\ \Delta V_2^{k+1} \\ \Delta\lambda^{k+1} \end{bmatrix}
$$

The final converged solution for a given tolerance (10^{-5}) is $[\delta^1 \quad V_2^1 \quad \lambda^1]^T = [-0.0129 \quad 1.0002 \quad 0.3]^T$. Then we can use this value as the starting point for the predictor and start the next step and so on.

After we get the tangent vector, we need to verify whether the system has reached the critical point. The sign of the product $dVd\lambda$ provides the information related to the critical point ($d\lambda=0$ corresponds to the critical point. If the sign of the product $dvd\lambda$ is positive then the critical point has been passed).

The tracing process based on continuation method includes the following three situations:

a) Tracing the upper part of the PV curve
b) Tracing near the critical point
c) Tracing the lower part of the PV curve

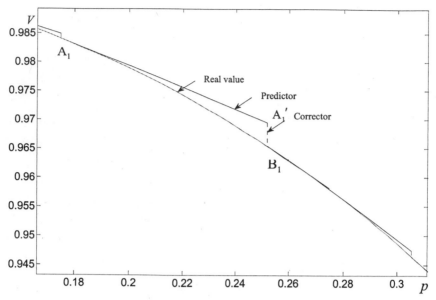

Fig.3.2 Tracing the upper part of the PV curve

a) Tracing the upper part of the PV curve:

As shown in Fig.3.2, suppose the predictor begins at A_1 (0.1750, 0.9841) ($p = PX/E^2$). At point A_1, the augmented Jacobian matrix is

$$J_{Aug} = \begin{bmatrix} 9.6837 & -1.7783 & 0.1 \\ -1.75 & 9.8406 & 0 \\ 0 & 0 & 1 \end{bmatrix}$$

So the tangent vector is

$$\begin{bmatrix} d\delta & dV_2 & d\lambda \end{bmatrix}^T = \begin{bmatrix} -0.0107 & -0.0019 & 1 \end{bmatrix}^T$$

The estimated step length σ at this point is 7.68.   Then the predictor becomes

$$\begin{bmatrix} \overline{\delta}^{k+1} \\ \overline{V}_2^{k+1} \\ \overline{\lambda}^{k+1} \end{bmatrix} = \begin{bmatrix} \delta^k \\ V_2^k \\ \lambda^k \end{bmatrix} + \sigma \begin{bmatrix} d\delta^k \\ dV_2^k \\ d\lambda^k \end{bmatrix} = \begin{bmatrix} -0.1788 \\ 0.9841 \\ 16.5 \end{bmatrix} + 7.68* \begin{bmatrix} -0.0107 \\ -0.0019 \\ 1 \end{bmatrix} = \begin{bmatrix} -0.2608 \\ 0.9695 \\ 24.18 \end{bmatrix}$$

which corresponds to the point A_1' (0.2518, 0.9695) in Fig.3.2. With same augmented Jacobian, one can perform corrector iterations using A_1' as initial guess for Newton method. After checking for proper convergence tolerance, the final solution is:

$$\begin{bmatrix} \delta^{k+1} \\ V_2^{k+1} \\ \lambda^{k+1} \end{bmatrix} = \begin{bmatrix} \overline{\delta}^{k+1} \\ \overline{V}_2^{k+1} \\ \overline{\lambda}^{k+1} \end{bmatrix} + \begin{bmatrix} \Delta\delta^{k+1} \\ \Delta V_2^{k+1} \\ \Delta\lambda^{k+1} \end{bmatrix} = \begin{bmatrix} -0.2639 \\ 0.9654 \\ 24.18 \end{bmatrix}$$

which is the point B_1(0.2518, 0.9654). Then we can begin the next predictor.

b) Tracing near the critical point

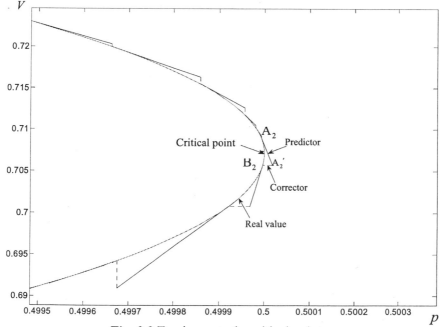

Fig. 3.3 Tracing near the critical point

As shown in Fig.3.3, suppose the predictor begins at A_2 (0.5, 0.709) which is close to the critical point.   At point A_2, the augmented Jacobian matrix is

$$J_{Aug} = \begin{bmatrix} 5.0264 & -7.0524 & 0.1 \\ -4.9999 & 7.0897 & 0 \\ 0 & 0 & 1 \end{bmatrix}$$

So the tangent vector is

$$\begin{bmatrix} d\delta & dV_2 & d\lambda \end{bmatrix}^T = \begin{bmatrix} -1.893 & -1.335 & 1 \end{bmatrix}^T$$

which corresponds to the point A_2' (0.5, 0.7057) in Fig.3.3. It should be noted here that the absolute value of $d\lambda$ is less than the other tangent vector absolute values, so we changed the continuation parameter from λ to V_2 for the corrector. For the corrector convergence, the step length σ is reduced to 0.0025.

The final corrected solution:

$$\begin{bmatrix} \delta^{k+1} \\ V_2^{k+1} \\ \lambda^{k+1} \end{bmatrix} = \begin{bmatrix} \overline{\delta}^{k+1} \\ \overline{V}_2^{k+1} \\ \overline{\lambda}^{k+1} \end{bmatrix} + \begin{bmatrix} \Delta\delta^{k+1} \\ \Delta V_2^{k+1} \\ \Delta\lambda^{k+1} \end{bmatrix} = \begin{bmatrix} -0.7874 \\ 0.7057 \\ 48.9996 \end{bmatrix}$$

which is the point B_2 (0.5, 0.7057).

Actually the real critical point of the system is (0.5, 0.7071), which is very close to B_2.

c) Tracing below the critical point

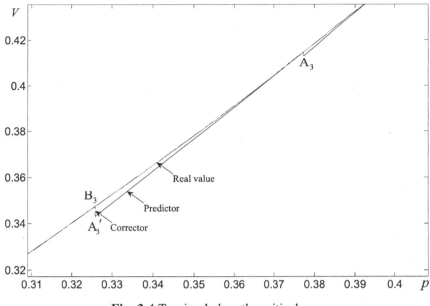

Fig. 3.4 Tracing below the critical

In Fig.3.4, suppose the predictor begins at A_3 (0.3774, 0.4147). At point A_3, the augmented Jacobian matrix is

$$J_{Aug} = \begin{bmatrix} 1.72 & -9.0995 & 0.1 \\ -3.7738 & 4.1473 & 0 \\ 0 & 0 & 1 \end{bmatrix}$$

So the tangent vector is

$$\begin{bmatrix} d\delta & dV_2 & d\lambda \end{bmatrix}^T = \begin{bmatrix} -0.0152 & -0.0139 & -1 \end{bmatrix}^T$$

The predictor then becomes ($\sigma = 5.154$)

$$\begin{bmatrix} \overline{\delta}^{k+1} \\ \overline{V}_2^{k+1} \\ \overline{\lambda}^{k+1} \end{bmatrix} = \begin{bmatrix} \delta^k \\ V_2^k \\ \lambda^k \end{bmatrix} + \sigma \begin{bmatrix} d\delta^k \\ dV_2^k \\ d\lambda^k \end{bmatrix} = \begin{bmatrix} -1.1432 \\ 0.4147 \\ 36.7379 \end{bmatrix} + 5.154* \begin{bmatrix} -0.0152 \\ -0.0139 \\ -1 \end{bmatrix} = \begin{bmatrix} -1.2217 \\ 0.3432 \\ 31.5839 \end{bmatrix}$$

which to the point A_3' (0.3258, 0.3432) in Fig.3.4. With same augmented Jacobian, one can perform corrector iterations using A_3' as initial guess for Newton method. After checking for proper convergence tolerance, the final solution is:

$$\begin{bmatrix} \delta^{k+1} \\ V_2^{k+1} \\ \lambda^{k+1} \end{bmatrix} = \begin{bmatrix} \overline{\delta}^{k+1} \\ \overline{V}_2^{k+1} \\ \overline{\lambda}^{k+1} \end{bmatrix} + \begin{bmatrix} \Delta\delta^{k+1} \\ \Delta V_2^{k+1} \\ \Delta\lambda^{k+1} \end{bmatrix} = \begin{bmatrix} -1.2159 \\ 0.3475 \\ 31.5839 \end{bmatrix}$$

which is the point B_3 (0.3258, 0.3475). Then we can begin the next predictor.

Finally, the whole tracing trajectory is shown in Fig.3.5. Compared with the PV curve for the example in Chapter 1, we can see that the results are identical. The PV curve in Fig.3.5 corresponds to unity power factor case of Fig.1.6 in Chapter 1.

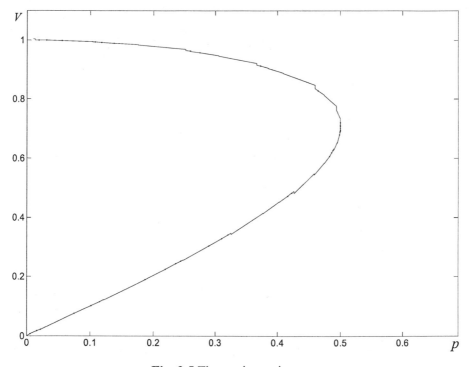

Fig. 3.5 The tracing trajectory

Two bus example: Nonlinear load:

The above two bus example is based on a constant power load model, and load is independent of voltage. Thus, load is made to vary in direct proportion to any change in λ. But in a nonlinear load model, the response of the load to a change in voltage magnitude must be considered, and in such load model, the load is not in direct proportion to change in λ. We can represent load increase scenario at bus 2 with nonlinear model as follows:

$$P = P_0(V_2/V_{20})^{KPV} * (1+\lambda K)$$
$$Q = Q_0(V_2/V_{20})^{KQV} * (1+\lambda K)$$

In the above equation, V_{20} is the initial voltage magnitude and parameter KPV and KPQ can be used to represent different load models. For example, $KPV = 0$, $KQV = 0$ is constant power model; $KPV = 1$, $KQV = 1$ is constant current model; $KPV = 2$, $KQV = 2$ is constant impedance model. With nonlinear load model, we can get new equations for real and reactive power:

$$0 = P_0(V_2/V_{20})^{KPV} * (1+\lambda K) + Y_{21}V_2\cos(\theta_{21} - \delta) + Y_{22}V_2^2\cos(\theta_{22}) = f_1(\delta, V_2, \lambda)$$

$$0 = Q_0(V_2/V_{20})^{KQV} * (1+\lambda K) - Y_{21}V_2\sin(\theta_{21} - \delta) - Y_{22}V_2^2\sin(\theta_{22}) = f_2(\delta, V_2, \lambda)$$

We use constant current and constant impedance load model to demonstrate the effect of different load model on critical point with 2 bus system. In the simulation, the system parameters are given as follows: $K = 1.0$, $P_0 = 0.14$, $\cos\psi = 1.0$, $Y_{12} = Y_{22} = 10$, $\theta_{12} = 90°$, $\theta_{22} = -90°$.

Figs 3.6, 3.7 and 3.8 show PV curve, P-λ, and V-λ for constant current load model. As parameter λ increases, power consumption first increases and reaches maximum and then decreases. Here λ_{max} does not correspond to P_{max}. Here λ can be interpreted as connected load as opposed to actual load.

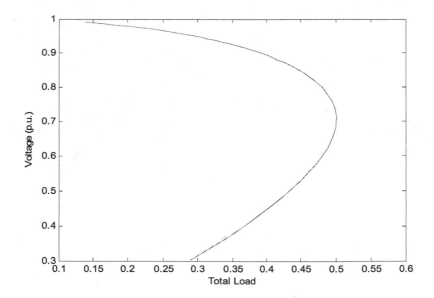

Fig. 3.6   V vs. P curve for constant current load

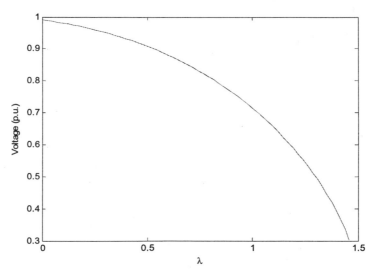

Fig. 3.7   V vs. λ curve for constant current load

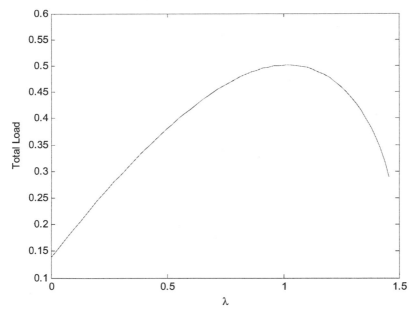

Fig. 3.8   P vs. λ for constant current load

Figs. 3.9, 3.10 and 3.11 show PV curve, P- λ , and V- λ for constant impedance load.

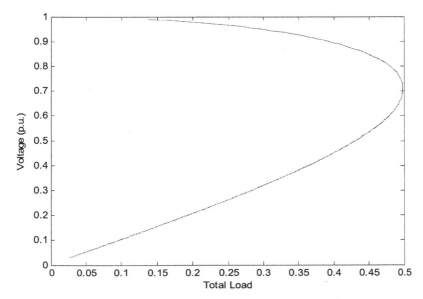

Fig. 3.9   V vs. P curve for constant impedance load

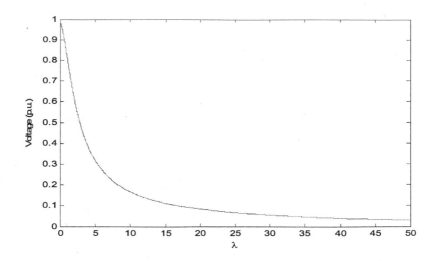

Fig. 3.10   V vs. λ curve for constant current load

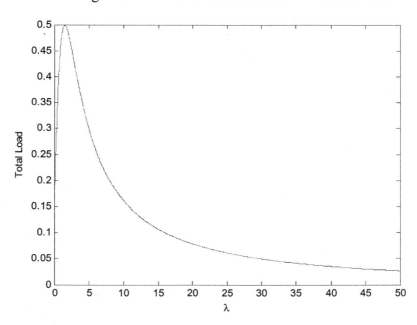

Fig. 3.11   P vs. λ for constant current load

For example if it is purely a resistive load then λ represents the number of parallel resistors connected at bus2. As we can see form the Fig. 3.11 , P_{max} occurs when the reactance of the transmission line(here we neglected the resistance of the transmission line) equals the total resistance connected at bus2. As the number of resistors increase beyond the one correspond to the maximum power, net load power decreases. The voltage at bus 2 also decreases with λ as shown in Fig. 3.10.

39 bus New England test system example

The data related to this test system is given in Appendix A: three scenarios are considered to demonstrate the capability of the continuation power flow.

Scenario 1

In the scenario 1, loads at 8 buses are increased, while the increased load is picked up by 9 generators. These load buses are bus 7, 8, 15, 16, 18, 20, 21, 23, and the load is increased proportional to their initial load levels. Scheduled generator buses are generators 30, 31, 32, 33, 34, 35, 36, 37, 38. The generator output is also increased proportional to their initial generations. Among these 9 generators, generator 31 is chosen as the slack bus. Besides scheduled generation output increment, generator 31 is also responsible for the load balance of the network.

For the load, the load increment is defined as:

$$P_{Li}(\lambda) = P_{Li0}[1 + \lambda K_{Li}]$$
$$Q_{Li}(\lambda) = P_{Li0} \tan(\Psi i)[1 + \lambda K_{Li}]$$
$$K_{Li} = P_{Total} / \sum P_{Li0}$$

The initial total load (P_{Total0}) is 6141MW with $\lambda = 0$, and in the next step, the total load is 6783MW with $\lambda = (6783 - 6141)/6141 = 0.104548$. The total load increment is 642MW which is distributed among 8 load buses proportional to their initial bus load. The following Table 3.1 shows the initial bus load level, the coefficients K_L, bus load increment, new load level and the power factor at each bus:

Table 3.1   Scenario 1 for load variation

Bus	Initial Load (MW)	K	Load Increment (MW)	New Load Level (MW)	Power Factor
7	233.8	2.221326	54.15286	287.9529	0.94 Lagging
8	522	2.221326	120.9059	642.9059	0.95 Lagging
15	320	2.221326	74.11854	394.1185	0.90 Lagging
16	329.4	2.221326	76.29577	405.6958	0.93 Lagging
18	158	2.221326	36.59603	194.596	0.98 Lagging
20	680	2.221326	157.5019	837.5019	0.99 Lagging
21	274	2.221326	63.464	337.464	0.92 Lagging
23	247.5	2.221326	57.32606	304.8261	0.95 Lagging

For the generator

$$P_{Gi} = P_{Gi0}(1 + \lambda K_{Gi})$$
$$K_{Gi} = P_{Gi0} / \sum P_{Gi0}$$

The following Table 3.2 shows the initial generator output, the coefficients K_G, generation increment and new generation level:

Table 3.2   Scenario 1 for generation distribution

Generators	Initial Generation (MW)	K_G	Generation Increment (MW)	New Generation (MW)
30	230	1.185	28.41914	258.4191
31	722.53	1.185	89.27687	811.8069
32	630	1.185	77.84372	707.8437
33	612	1.185	75.61962	687.6196
34	488	1.185	60.29799	548.298
35	630	1.185	77.84372	707.8437
36	540	1.185	66.72319	606.7232
37	520	1.185	64.25196	584.252
38	810	1.185	100.0848	910.0848

Because generator 31 is the slack bus, the actual output is 827MW instead of scheduled 811MW, and the additional 16MW is for the load balance.

Fig.3.12 shows the variation of individual generators real power with total system load. Fig.3.13 shows the variation of individual bus load with respect to the total system load. Fig.3.14 presents PV curves at four critical load buses (these buses are based on the largest tangent vector elements corresponding to voltages at the critical point). Fig.3.15 shows the relationship between the bus voltage and the parameter λ. For constant power loads λ_{max} corresponds to P_{max}. However for nonlinear loads λ_{max} does not correspond to P_{max} as shown in two bus example.

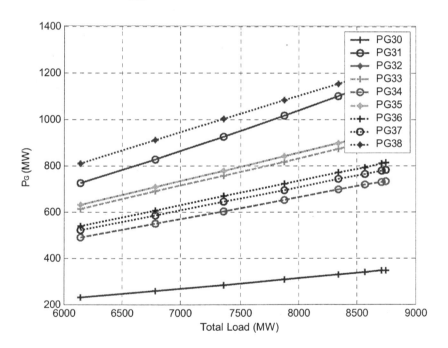

Fig.3.12 P_G vs. P_{total}

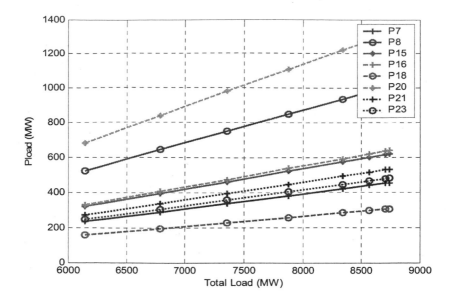

Fig.3.13 P_{load} at bus vs. P_{total}

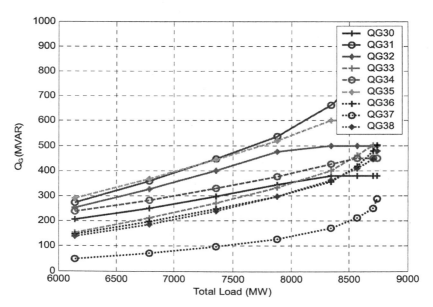

Fig.3.14. Q_g vs. P_{total}

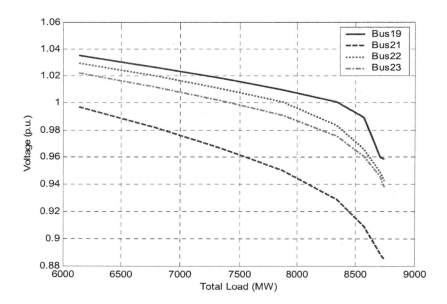

Fig.3.15   V vs. P$_{total}$

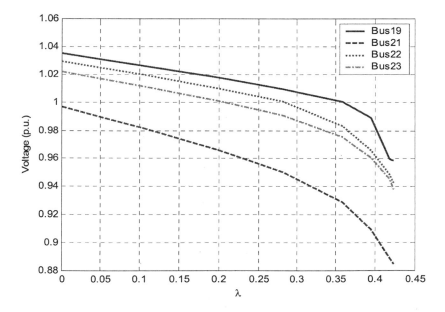

Fig.3.16   V vs. λ

Scenario 2

The only difference in this scenario compared to scenario 1 is that the loads at those 8 buses are increased equally. This means load increment is the same among all the load buses. All other conditions are same as in scenario1.

Table 3.3   Scenario 2 for load variation

Bus	Initial Load (MW)	K	Load Increment (MW)	New Load Level (MW)
7	233.8	3.283136	79.16	312.96
8	522	1.470493	79.17	601.17
15	320	2.398741	79.16	399.16
16	329.4	2.330289	79.17	408.57
18	158	4.85821	79.16	237.16
20	680	1.128819	79.18	759.18
21	274	2.80145	79.17	353.17
23	247.5	3.101403	79.15	326.65

Figs.3.17 to 3.21 are similar to Fig.3.12 to 3.16 respectively.

The initial total load is 6141MW with $\lambda = 0$, and in the next step, the total load is 6774MW with $\lambda = (6774 - 6141)/6141 = 0.103127$. The total load increment is 633MW which is distributed among 8 load buses equally. The following Table 3.3 shows the initial bus load level, the coefficients K_L, bus load increment and new bus load level. The power factor at each bus is same as in scenario1.

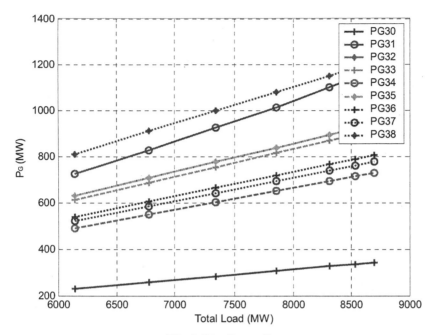

Fig.3.17 P_g vs. P_{total}

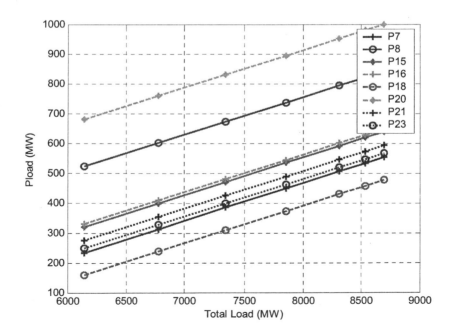

Fig 3.18 P_{load} at various buses vs. P_{total}

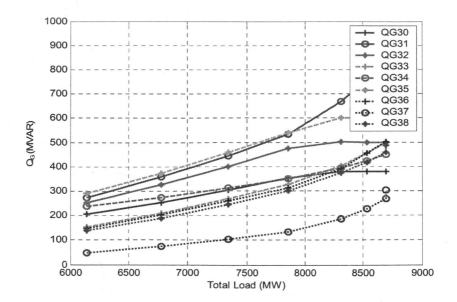

Fig. 3.19 Q_G vs. P_{total}

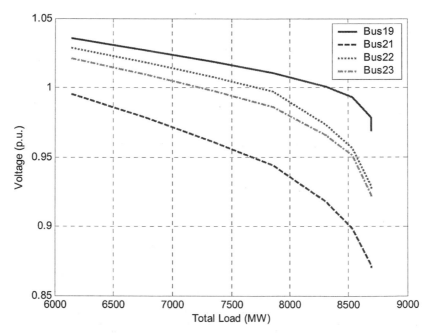

Fig.3.20   V vs. P_{total}

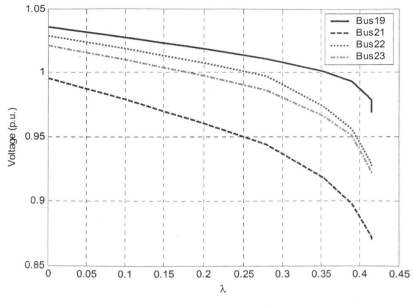

Fig 3.21   V vs. λ

3.6 Simultaneous Equilibria Tracing in Power Systems

In power system analysis, it is frequently of interest to find solutions of the system at an equilibrium point. For instance, the solution of the power flow equations is needed in system planning and static security analysis. In stability analysis, a power flow is used to calculate the voltages and angles at all buses, and then the dynamic state variables are evaluated using the device equations. This procedure causes some problems as will be shown in the following sections. To overcome these problems, we will further extend the continuation technique to simultaneously trace the total system equlibria of the structure preserving power system model, which is described by a set of nonlinear differential and algebraic equations (DAEs). Physical interpretations of the new approach will give insights into some issues which are important to a good understanding of the power system.

Unlike in power flow analysis, a detailed dynamic representation of the power system is required to analyze the system's stability behavior. As a typical nonlinear dynamic system, with the multiple time-scale property, a set of nonlinear DAEs can be employed to describe the behavior of the power system, i.e.,

$$\dot{X} = F(X, Y, P) \qquad\qquad (3.13)$$

$$0 = G(X, Y, P) \qquad\qquad (3.14)$$

where X includes the dynamic states, Y includes the algebraic state variables, and P consists of all parameters explicitly appearing in F and G. Some of these parameters can be control input settings.

3.6.1 Total solution at an equilibrium

A system equilibrium solution is needed for the evaluation of the stability, the solution X_0 and Y_0 of Eqs.3.13 and 3.14 at steady state, i.e., when $\dot{X} = 0$, constitute the equilibrium point. Setting the differential to zero indicates a state of equilibrium of the system. In small signal stability analysis, the right hand side of the DAEs is first linearized, and then the system state matrix A_{sys} (where $A_{sys} = F_X - F_Y G_Y^{-1} G_X$, see section 3.8.7) is evaluated at (X_0, Y_0). Its eigenvalues give small signal stability information of the current equilibrium point. In nonlinear time domain analysis, the equilibrium solution (X_0, Y_0) gives the initial conditions to start numerical integration. In direct Lyapunov type stability analysis, this solution is also required.

3.6.2 Traditional approach

In Eq.3.14, G corresponds to the power balance equations at all buses in the system. Therefore its dimension is larger than that of the power flow. In power flow, it is assumed that the voltage at PV buses and voltage and angle at the slack bus(es) are known and constant. Consequently, for a network of n buses, if there are m generators, m_s of which are designated as slack, then the number of equations in the power flow formulation will be $2n - m - m_s$ (for polar coordinates). For a constant generator terminal voltage, it is assumed that the static gain of the excitation system is infinite. No limitations on the slack bus generation can be enforced during the solution process. Once a power flow is solved, together with the pre-specified generation and voltages for PV and slack buses, the X_0

values will be updated using Eq.3.13 at steady state. The control parameter settings in P corresponding to this X_0 are then computed. This procedure of solving for (X_0, Y_0, P_0) is termed as the two-step approach. With this total system equilibrium solution, further stability analysis can then be conducted.

The above procedure has some drawbacks. Firstly, if control limits are enforced, a solution (X_0, Y_0, P_0) satisfying these limits may not exist. The slack bus generation might also exceed limits after the power flow. In this case, the state which is limited would need to be fixed at its limiting value and a corresponding new steady state equilibrium solution would have to be found. This would require a new power flow, for each specified value of PV bus generation or terminal voltage, or possibly generator reactive power injection. For the last case, the generator voltage becomes part of the power flow solution. For a heavily loaded system, this trial and error procedure may have to be repeated several times, each time requiring a new power flow solution. Secondly, even after a set of (X_0, Y_0, P_0) values satisfying all limits are found, there still exists another problem which is inevitable in using the power flow based two-step approach to produce equilibria solution for stability analysis. That is, the description of the generators in the power flow is very different from that in the dynamic response. How the generators behave in a dynamic process depends on the dynamic characteristics of the synchronous machine and the control systems associated with them. These controls are not represented for the PV bus generators and the slack bus generators are simply left out in the power flow. Therefore, it may not be unusual that this discrepancy in representation leads to erroneous results.

3.7 Power Flow Methodology and Assumptions

Before introducing the simultaneous equilibria tracing technique, let us first have a closer look at the assumptions used in the power flow, particularly the reasons why they are needed. With a clearer understanding of these assumptions, we will then be able to devise a procedure in which the problems encountered in the traditional approach can be avoided.

3.7.1 Nonlinearity in power flow

In normal electrical network analysis, the voltages and/or currents of power sources are given as known quantities. In order to find the voltages at various nodes and currents in all branches, one simply needs to solve the network nodal equations which are linear. Correspondingly, for power network, this refers to the nodal representation, given in phasor notation as

$$Y\overline{V} = \overline{I} \tag{3.15}$$

where Y is the network admittance matrix, $\overline{V}$ is the vector of phasor voltages at all buses, and $\overline{I}$ is the nodal phasor injection currents. The conditions for Eq.3.15 to have a solution with a specified set of injection currents $\overline{I}$ are

- Y is nonsingular;

- $\mathrm{rank}(Y\,|\,\overline{I}) = \mathrm{rank}(Y)$ if Y is singular.

Were the injection currents known, the power flow would have involved no nonlinear equations. However, in power system analysis, the nodal voltages and injection currents are both unknown before a power flow is solved. Instead, the generation and load powers are given as the known quantities. They are related to the nodal voltages and injection currents as shown below.

$$\overline{I}_i = \frac{S_{Gi} - S_{Li}}{\overline{V}_i^*}$$

$$= \frac{(P_{Gi} - P_{Li}) + j(Q_{Gi} - Q_{Li})}{\overline{V}_i^*}$$

The '*' sign indicates the complex conjugate. With the real and imaginary parts separated, Eq.3.15 is transformed into the following form

$$0 = P_{Ei} - P_{Li} - P_{Ti} \qquad i = 1, \cdots n \tag{3.16}$$

$$0 = Q_{Ei} - Q_{Li} - Q_{Ti} \qquad i = 1, \cdots n \tag{3.17}$$

where

$$P_{Ti} = V_i \sum_{k=1}^{n} V_k y_{ik} \cos(\theta_i - \theta_k - \gamma_{ik}) \qquad (3.18)$$

$$Q_{Ti} = V_i \sum_{k=1}^{n} V_k y_{ik} \sin(\theta_i - \theta_k - \gamma_{ik}) \qquad (3.19)$$

The two nonlinear Eqs.3.16 and 3.17 correspond to the algebraic part of the DAE formulation given in Eq.3.14. With the powers specified at the terminal buses, X variables are not of concern in the power flow equations.

3.7.2 Slack bus assumption

The unknowns in Eqs.3.16 and 3.17 are $(\underline{V}, \underline{\theta})$, the number of which is $2n$. The underline sign is used to denote vectors. If we want o solve these unknowns directly using the Newton's method, we have to specify the generations and load powers at all buses. And most probably, with a starting point $(\underline{V}^0, \underline{\theta}^0)$ close to normal operating conditions, this approach will lead to divergence. A closer look of the structure of the power balance equations will give more insight into the problem. Designating the generator at the nth bus as the slack, summing up the first $n-1$ equations in Eqs.3.16 and 3.17 and then adding them to the nth and $2n^{th}$ equations respectively will yield

$$P_{Gs} = \sum_{i=1}^{n} P_{Li} + P_{loss}(\underline{V}, \underline{\theta}) - \sum_{i=1}^{n-1} P_{Ei} \qquad (3.20)$$

$$Q_{Gs} = \sum_{i=1}^{n} Q_{Li} + Q_{loss}(\underline{V}, \underline{\theta}) - \sum_{i=1}^{n-1} Q_{Ei} \qquad (3.21)$$

Since we know that

$$\sum_{i=1}^{n} P_{Ti} = P_{loss}(\underline{V}, \underline{\theta})$$

$$\sum_{i=1}^{n} Q_{Ti} = Q_{loss}(\underline{V}, \underline{\theta})$$

These two equations can be put together with the first $n-1$ equations from 3.16 and 3.17 respectively to represent the complete network. Eqs. 3.20 and 3.21 show that, if a solution $(\underline{V}^*, \underline{\theta}^*)$ exists, for a possible successful convergence, we must specify the generations subject to the constraints given in Eqs.3.20 and 3.21. Since the losses as a function of the network solution are unknown before the power flow is solved, it is practically impossible to do so. Therefore, it is very likely that, if we have to specify the power generations for all generators, constraints Eqs.3.20 and 3.21 may be greatly violated, and correspondingly the staring point $(\underline{V}^0, \underline{\theta}^0)$ might be well out of the radius of convergence of the Newton's method. Also, there is a possibility that a *real* solution simply does not exist corresponding to this set of specified generations. (From algebraic equations theory, we know that a solution always exists if we also consider complex roots.) If one can devise a scheme so that there is freedom of adjusting the generation during the course of iteratively solving the power flow equations, then convergence performance might be much better. Referring to this, an immediate thought would be to eliminate Eqs.3.20 and 3.21 altogether from the power flow iterations. Consequently, the slack bus generation need not be specified. To do so, we must remove two unknowns from $(\underline{V}, \underline{\theta})$. This is no difficulty at all. Because the goal of a power flow is to give a dispatch of the generation so that the system load can be served with the bus voltages being close to normal operating conditions, usually close to 1.0 p.u., we can reasonably assign 1.0 to V_s and 0^0 to θ_s, the latter of which is simply to set a reference for the angle measurement, and thus it is arbitrary. After the power flow converges, we then calculate the losses and assign all of them to the slack generators. This procedure makes sure that the loss-generation imbalance does not cause convergence trouble during iteration. And this imbalance is fixed *only* after the power flow is solved. The above discussion shows that the slack bus assumption is a mathematical requirement for possible/good convergence of the Newton's iterative algorithm.

3.7.3 PV bus assumption

In order to maintain the system voltage levels, the generators are equipped with automatic voltage regulators (AVR) so that terminal voltages are within limits during system load increase or other disturbances. With the power flow description of the system, the only way to reflect this fact is to force the terminal voltages at the generator buses as constant since AVR is not represented. To achieve this, the reactive power balance equations for generator buses must be removed. As a consequence, Q_{Ei} is no longer needs to be specified as input, it is released as a variable. Physically, this means that reactive support from generators helps maintain a relatively high and steady terminal voltage. Numerically, this possibly also leads to better convergence characteristics of the Newton-Raphson power flow algorithm.

After the above discussion, we are now ready to devise a strategy that eliminates the unreasonable assumptions used in the power flow. It solves for a reasonable set of (X, Y) values with control limits automatically implemented. This leads us to the topic of simultaneous equilibria tracing technique.

3.8 Total Power System Equilibria Solutions

From the discussion given in Section 3.7 , we can make two conclusions about the assumptions used in the power flow:

- *Slack bus methodology* provides a means of "automatically" adjusting real and reactive power generation "during" the iteration, not at all buses, but only for the slack, so that at any iteration the losses are not causing the point to be too far away from the true solution, therefore making Newton's iterative method possible to converge.

- *The PV bus assumption* is used to reflect the need of maintaining the system voltage levels by AVRs and it also possibly helps improve the convergence rate of the Newton-Raphson algorithm.

In the following sections, we will study how these assumptions, which cause the problems mentioned in subsection 3.6.2 can be removed, while the goals they are made to achieve are not sacrificed.

Before we introduce the simultaneous equilibria tracing technique, let us first give a detailed representation of the structure-preserving power system model.

3.8.1 Formulation of power system DAE model

A power system is assumed to have n buses and m generators. Each generator is assumed to be equipped with the same type of excitation control system and speed governor. The formulation of power system modeling is presented in this chapter. The commonly used power system notations are adopted here.

3.8.1.1 Synchronous generators

Without loss of generality, the rotor angle of the mth generator is chosen as the system angle reference. This choice of reference is different from the conventional slack bus selection. No assumptions are necessary for choosing such a reference. When stator transients are ignored, the two-axis model [5, 6] describing the synchronous machine dynamics can be given as:

$$\dot{\delta}_i = (\omega_i - \omega_m)\omega_0 \quad i = 1, \cdots, m-1$$
(3.22)

$$\dot{\omega}_i = M_i^{-1}[P_{mi} - D_i(\omega_i - \omega_m) - (E_{qi}' - X_{di}'I_{di})I_{qi}$$
$$- (E_{di}' + X_{qi}'I_{qi})I_{di}] \quad i = 1, \cdots, m$$
(3.23)

$$\dot{E}_{qi}' = T_{d0i}^{-1}[E_{fdi} - E_{qi}' - (X_{di} - X_{di}')I_{di}] \quad i = 1, \cdots, m$$
(3.24)

$$\dot{E}_{di}' = T_{q0i}^{-1}[-E_{di}' + (X_{qi} - X_{qi}')I_{qi}] \quad i = 1, \cdots, m$$
(3.25)

where ω_m is the system frequency, ω_i is the machine frequency, namely, generator angular speed and ω_0 is the system rated frequency (377.0 rad/sec). I_{di} and I_{qi} are direct axis and quadrature axis currents respectively; E_{di}' and E_{qi}' are transient direct axis and quadrature axis EMF respectively; T_{d0i} and T_{q0i} are direct axis and quadrature axis open circuit time constants respectively; X_{di}' and X_{qi}' are direct axis and quadrature axis transient reactances and R_{si} are armature resistance of the machine; M_i is inertia constant and D_i is the damping constant of the machine. All the quantities are in per unit except ω_0.

Interface voltage equations to the network are given as follows:

$$E'_{qi} = V_i \cos(\delta_i - \theta_i) + R_{si} I_{qi} + X'_{di} I_{di} \tag{3.26}$$

$$E'_{di} = V_i \sin(\delta_i - \theta_i) + R_{si} I_{di} - X'_{qi} I_{qi} \tag{3.27}$$

where V_i and θ_i are bus voltage and angle respectively.

The machine currents I_{di} and I_{qi} can be eliminated by solving the generator interface equations to the network. Hence,

$$I_{di} = [R_{si}E'_{di} + E'_{qi}X'_{qi} - R_{si}V_i \sin(\delta_i - \theta_i) - X'_{qi}V_i \cos(\delta_i - \theta_i)]A_i^{-1} \tag{3.28}$$

$$I_{qi} = [R_{si}E'_{qi} - E'_{di}X'_{di} - R_{si}V_i \cos(\delta_i - \theta_i) - X'_{di}V_i \sin(\delta_i - \theta_i)]A_i^{-1} \tag{3.29}$$

$$A_i = R_{si}^2 + X'_{di}X'_{qi} \tag{3.30}$$

Note that Eq.3.22 does not include the differential equation for δ_m, and all the angles here and henceforth are relative angles with respect to the mth generator's rotor angle.

3.8.1.2 Excitation Control system

The simplified IEEE type DC-1 excitation system [5] as shown in Fig.3.22 is used here. The corresponding mathematical model is

$$\dot{E}_{fdi} = T_{ei}^{-1}[V_{ri} - [S_{ei}(E_{fdi})]E_{fdi}] \quad i = 1, \cdots, m \tag{3.31}$$

$$\dot{V}_{ri} = T_{ai}^{-1}[-V_{ri} + K_{ai}(V_{refi} - V_i - R_{fi})] \quad i = 1, \cdots, m \tag{3.32}$$

If $V_{ri,min} \le V_{ri} \le V_{ri,max}$, $V_{pssi} = 0$ (at steady state),

$$\dot{R}_{fi} = T_{fi}^{-1}[-R_{fi} - [K_{ei} + S_{ei}(E_{fdi})]K_{fi}E_{fdi}/T_{ei} + K_{fi}V_{ri}/T_{ei} \qquad (3.33)$$

where V_{ref} is the reference voltage of the automatic voltage regulator (AVR); V_{ri} and R_{fi} are the outputs of the AVR and exciter soft feedback; E_{fdi} is the voltage applied to generator field winding; T_{ai}, T_{ei} and T_{fi} are AVR, exciter and feedback time constants respectively ; K_{ai}, K_{ei} and K_{fi} are gains of AVR, exciter and feedback respectively; $V_{ri,min}$ and $V_{ri,max}$ are the lower and upper limits of V_{ri} respectively.

3.8.1.3 Prime mover and speed governor

Fig.3.23 shows the block diagram for a simplified prime mover and speed governor. Two differential equations are involved to describe the dynamics when no μ_i limit is hit.

$$\dot{P}_{mi} = T_{chi}^{-1}(\mu_i - P_{mi}) \qquad i = 1,\cdots,m \qquad (3.34)$$

$$\dot{\mu}_i = T_{gi}^{-1}[P_{gsi}(\omega_i - \omega_{ref})/R_i - \mu_i] \qquad \text{if } \mu_{i,min} \le \mu_i \le \mu_{i,max} \qquad (3.35)$$
$$i = 1,\cdots,m$$

where $P_{gsi} = P_{gsi}^0(1 + K_{gi}\mu)$ is the designated real power generation; P_{gsi}^0 is its setting at base case; K_{gi} is the generator load pick-up factor that could be determined by AGC, EDC or other system operating practice; P_{mi} is the mechanical power of prime mover and μ_i is the steam valve or water gate opening; R_i is the governor regulation constant representing its inherent speed-droop characteristic; $\omega_{ref}(=1.0)$ is the governor reference speed; T_{chi} and T_{gi} are the time constants related to the prime mover and speed governor respectively; $\mu_{i,min}$ and $\mu_{i,max}$ are the lower and upper limits of μ, where a parameter μ is introduced to designate the system operation scenario. At the base case, μ equals to zero.

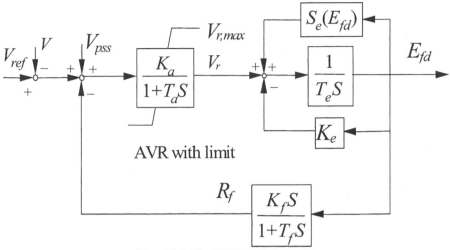

Fig. 3.22 The IEEE DC-1 model

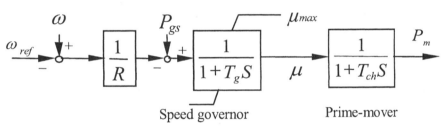

Fig. 3.23   The simplified speed governor and prime mover

3.8.1.4 Nonlinear load model

The voltage and frequency dependent load can be modeled as follows:

$$\begin{cases} P_{li} = P_{li0}(V_i / V_{i0})^{\alpha_i}[1 + K_{lpfi}(\omega_m - \omega_r)] \\ Q_{li} = Q_{li0}(V_i / V_{i0})^{\beta_i}[1 + K_{lqfi}(\omega_m - \omega_r)] \end{cases} \quad i = 1, \cdots, m \quad (3.36)$$

where P_{li0} and Q_{li0} are the active and reactive powers absorbed by the load at the nominal voltage V_i and frequency $\omega_r (=1.0)$. The frequency dependent term is included to prevent the equilibrium computation from divergence in case all the generators reach their maximum real power limits due to load increase or generator outages. Here K_{lpf} and K_{lqf} are the load changing factors with respect to system frequency.

3.8.1.5 LTC model

Eqs.3.37 and 3.38 show the response of Load Tap Changer modeled as continuous. Assume the transformer is between bus i and bus j, then

$$V_j = rV_i \tag{3.37}$$

$$T_t \dot{r} = V_j^{ref} - V_j \tag{3.38}$$

where r is the tap ratio of an LTC; V_j^{ref} is the reference voltage at the LTC regulated bus j; T_t is the time constant.

3.8.1.6 Other Models

Generic Dynamic Load Model: Ref [20] proposed a generic load model to capture nonlinear characteristics as well load recovery. This model includes both steady state and transient load characteristics. Ref [21] further classifies this model in term of multiplicative generic model and additive generic model based on how load state variables affect the transient load characteristic. For additive generic model the corresponding equations that represent relevant responses are [21]:

Transient load response at particular bus i:

$$P_{liT} = P_{li0}[z_{Pi} + (\frac{Vi}{V_{i0}})^{\alpha_T}] \tag{3.39}$$

$$Q_{liT} = Q_{li0}[z_{Qi} + (\frac{Vi}{V_{i0}})^{\beta_T}] \tag{3.40}$$

Where the dyanmic state variables Z_{Pi} and Z_{Qi} represented by the following equations:

$$T_{Pi} \frac{d}{dt} z_{Pi} = -z_{Pi} + \left(\frac{V_i}{V_{i0}}\right)^{\alpha_S} - \left(\frac{V_i}{V_{i0}}\right)^{\alpha_T} \tag{3.41}$$

$$T_{Qi} \frac{d}{dt} z_{Qi} = -z_{Qi} + \left(\frac{V_i}{V_{i0}}\right)^{\beta_S} - \left(\frac{V_i}{V_{i0}}\right)^{\beta_T} \tag{3.42}$$

At steady state

$$P_{liS} = P_{li0} \left(\frac{V_i}{V_{i0}}\right)^{\alpha_S} \tag{3.43}$$

$$Q_{liS} = Q_{li0} \left(\frac{V_i}{V_{i0}}\right)^{\beta_S} \tag{3.44}$$

These models can be easily incorporated in general DAE formulation for equilibrium tracing or for time domain simulation. Time domain simulations for short term and for long term are discussed in chapter 6. Ref [21] also includes one chapter that provides wide coverage of various load aspects including the induction motor for voltage stability studies.

HVDC Models:

HVDC can also be easily incorporated into power flow. References [5, 7] provide a systematic presentation for power flow formulation for AC-DC power systems. These references cover AC –DC power flow solution both for single converter and multiterminal DC systems. Basically, there are real and reactive power mismatches at the converter terminal bus bars similar to AC power mismatches at particular AC bus. At converter terminal bus bar there are additional real and reactive power injections which are function of DC system variables and the converter AC terminal bus bar voltage. The variable corresponds to DC include: average DC voltage, converter DC current, firing angle of the converter, the converter transformer off- nominal tap ratio [7].

Ref [23] specifically discusses the comparison between point of collapse methods and continuation methods for large scale AC/DC systems. The detailed modeling related to AC-DC dynamic systems for point of collapse methods is discussed in ref [24]. This reference included inverter and converter control functions and showed voltage dependent current order limits affect the voltage stability margin of the system. This reference also observed Hopf bifurcation. For Hopf bifurcation, voltage dependent current order limit also played an important role.

In a recent paper [27] a single-infeed HVDC model is incorporated in combination with detailed synchronous machine modeling and excitation voltage control. The authors derived analytical expressions for power/voltage stability indices.

FACTS Device Models:

Reactive power plays an important role in voltage stability studies. Flexible AC Transmission Systems (FACTS), such as Static Var Compensator (SVC), Thyristor Controlled Series Capacitor (TCSC), Static Synchronous Compensator (STATCOM), and Unified Power Flow Controller (UPFC) can provide required fast control to improve voltage stability.

Reference [8] came with a general FACTS device model that is flexible enough to represent any FACTS device including the ones mentioned above. Functional characteristics of various FACTS devices are derived from Voltage Source Converter (VSC) model. This model can be used both for power flow as well stability studies. Ref [28] provided steady state models for SVC and TCSC with controls for voltage stability analysis.

3.8.1.7 Network power equations

Corresponding to the above models, the network equations can be written as:

$$\begin{cases} 0 = P_{gi} - (1 + K_{Li}\lambda)P_{li} - P_{ti} \\ 0 = Q_{gi} - (1 + K_{Li}\lambda)Q_{li} - Q_{ti} \end{cases} \quad i = 1, \cdots, n \qquad (3.45)$$

Where

$$\begin{cases} P_{ti} = \sum_{k=1}^{n} V_i V_k Y_{ik} \cos(\theta_i - \theta_k - \varphi_{ik}) \\ Q_{ti} = \sum_{k=1}^{n} V_i V_k Y_{ik} \sin(\theta_i - \theta_k - \varphi_{ik}) \end{cases} \quad i = 1, \cdots, n \qquad (3.46)$$

And

$$\begin{cases} P_{gi} = I_{di}V_i \sin(\delta_i - \theta_i) + I_{qi}V_i \cos(\delta_i - \theta_i) \\ Q_{gi} = I_{di}V_i \cos(\delta_i - \theta_i) - I_{qi}V_i \sin(\delta_i - \theta_i) \end{cases} \quad i = 1, \cdots, m \qquad (3.47)$$

P_{gi} and Q_{gi} are the generator output powers, which are primarily determined by the inherent characteristics of the speed governor and the AVR regulations. They will change if real power generation rescheduling and secondary voltage control is applied. P_{ti} and Q_{ti} are the powers injected into the network at bus i. K_{Li} is load changing factor specified for bus i as mentioned in Section 3.8.3.

3.8.1.8 Power system DAE model

The above differential and algebraic equations are commonly known as a DAE representation of the power system. In a compact form, they can be simply represented by Eqs.3.13 and 3.14 as described in Section 3.6.

The state vector X and algebraic vector Y contain the following variables:

$$X = (\delta, \omega, E_q', E_d', P_m, \mu, E_{fd}, V_r, R_f)$$
$$Y = (V, \theta) \tag{3.48}$$

The P in Eqs.3.13 and 3.14 can be further divided into control vector U and parameter vector Z

$$U = (V_{ref}, P_{gs}, \cdots), \quad Z = (P_l, Q_l) \tag{3.49}$$

In short, X contains all the system state variables; Y includes the algebraic variables; U is the control vector, whereas Z characterizes system loading condition.

3.8.2 Bifurcation modeling of power system dynamics

As discussed in Chapter 3, for a dynamic system parameterized by a single or a set of static parameters, bifurcations occur when the character of equilibrium changes within an arbitrary small local neighborhood of a critical parameter set. Those static parameters are defined as bifurcation parameters. Note that the prerequisite condition of bifurcation parameters is that their derivatives always equal zero. That is, they are out of dynamic variable set that characterize system state.

An extensive power system literature is available for the application of bifurcation related approach to voltage stability [9].

In power system DAE model, the change of equilibrium character with respect to bifurcation parameter is often effectively studied by analyzing changes of the eigenvalues of $A_{sys}(\lambda) = F_X - F_Y(G_Y)^{-1}G_X$ in response to parameter variations.

The various types of bifurcation points generally will form surfaces or manifold in the multidimensional parameter space. These surfaces serve in the parameter space as boundaries separating regions where a certain type of system operation (as characterized by equilibria and trajectories) persists. A point on such a surface can be identified by a single bifurcation parameter $\lambda = \lambda_0$. These bifurcations are classified as codimension one. Only local codimension one bifurcations are discussed here.

3.8.2.1 Saddle-node bifurcation

Saddle-node bifurcation occurs when the Jacobian of the system $A_{sys}(\lambda_0)$ has a simple eigenvalue and there is no other eigenvalue on the imaginary axis. The equilibrium cease to exist when λ moves beyond λ_0. Correspondingly in the state space x, two equilibriums approach each other as λ approaches λ_0; then at λ_0 they fuse in a nonhyperbolic equilibrium (with a zero eigenvalue).

Under certain additional transversality (non-degenerate) conditions, the presence of the simple zero eigenvalue of the Jacobian essentially characterizes this bifurcation. In second-order systems, this bifurcation corresponds to the annihilation of a saddle point and a node, hence the name saddle-node bifurcation [10].

3.8.2.2 Hopf bifurcation

When Hopf bifurcation occurs, the Jacobian A_{sys} of the system has a simple pair of purely imaginary eigenvalues and there are no other eigenvalues on the imaginary axis. As the parameter changes, certain inequality conditions need to hold that ensure that this pair of critical eigenvalues crosses the imaginary axis. They can be formulated as

$$\frac{\partial}{\partial \lambda} \text{Re}[\mu(\lambda)] \neq 0$$

when $\text{Re}(\mu)$ denotes the real part of the eigenvalue μ, which moves across the imaginary axis, and $\partial/\partial\lambda$ denotes the derivative with respect to the bifurcation parameter λ.

Typically, this means that for $\lambda \neq \lambda_0$ the system has an equilibrium and a closed trajectory; a limit cycle exists near this equilibrium for one side of the parameters. This limit cycle can be unstable (or stable), that is, trajectories diverge (converge) from (to) it from both the inside and the outside.

The stable limit cycle corresponds to super critical Hopf bifurcation. The unstable limit cycle corresponds to sub critical Hopf bifurcation. The supercritical Hopf bifurcation corresponds to a transition in the system operating condition from a small-signal stable equilibrium point for $\lambda < \lambda_c$, and a small-signal stable limit cycle for $\lambda > \lambda c$. That is, when the system undergoes a supercritical Hopf bifurcation at $\lambda = \lambda c$, the system operating condition changes to sustained oscillation for $\lambda > \lambda c$. This type of supercritical Hopf bifurcation appears and played a fundamental role in the oscillating event experienced by Union Electric in 1992 [11, 12]. Ref [12] applied Hopf bifurcation analysis to large scale power system. The authors in this reference studied power system model in DAE form. Hopf related segments are traced by continuation based approaches. The critical eigenvalue is estimated either by power method or modified Arnoldi method.

3.8.3 Manifold models in power systems

Mathematical models of many, practically important scientific and technical problems involve differentiable manifolds. Differentiable manifolds are implicitly defined as the solution sets of systems of nonlinear equations. The mathematical basis for manifold and its numerical treatment are well established in the mathematical literature [3,13,15]. Following sections provide brief summary based on these references.

3.8.3.1 Manifold

Assume a dynamic system is represented by:

$$\dot{x} = F(x,\lambda) \quad F : R^n \times R^d \to R^n \tag{3.50}$$

where F is a sufficiently smooth mapping, $x \in R^n$ is a state variable vector, $\lambda \in R^d$ is a parameter vector. The computational study of equilibria leads to nonlinear equations of 3.51.

$$F(x,\lambda) = 0 \tag{3.51}$$

As we see in the previous sections, it is of interest to determine the behavior of the solution under variation of λ. Here λ is a vector of parameters. The zero set $M = \{(x, \lambda) \in R^n \times R^d : F(x, \lambda) = 0\}$ has the structure of a submanifold of dimension d of the product $R^n \times R^d$ of state and parameter space. Computational techniques are well developed to find the critical point of interest on the manifold or any other dynamic behavior of interest. This approach can also be used in connection with equality constrained dynamical systems that are modeled by differential-algebraic equations (DAE) which are of interest to power system security analysis. Such DAE is known to be closely related to ordinary differential equations (ODE) on implicitly defined differentiable manifolds [13].

The basic computational problems arising in connection with any implicitly defined manifold is to come up with certain parameterizations. Finally it leads to solving certain set of nonlinear equations.

3.8.3.2 Natural parameterization

In many applications one can identify certain quantities that can be independently changed (for example constant load power change in power systems). This can be identified as parameter. This means that we have an intrinsic splitting, which includes a d-dimensional parameter space Λ and a state space X of dimension n.

$$X \oplus \Lambda \quad \text{and} \quad \dim(\Lambda) = d$$

This is a natural parameter splitting of original variable space. One can use the parameter space Λ as the coordinate space of a local coordinate system.

For example for power system we can identified parameters involved in power flow type formulation as well DAE type formulation. For some cases, the natural parameterization may be not suitable to be a local parameterization, in which cases singularity is always encountered while solving for the solution of nonlinear equation system.

3.8.3.3 Local parameterization

Rheinboldt [13] described mathematics behind the local parameterization to trace the equilibrium curve. The local parameterization could avoid the singularity encountered by the natural parameterization. The procedure for one parameter problem is described here. Continuation methods produce a sequence of solutions for changing parameter. Local parameterization provides a way to trace this path successfully.   Local parameterization at a given step requires a nonzero vector $e \in R^n$ such that

$$e \notin rangeDF\ (x)^T \qquad (3.52)$$

Basically (3.52) implies e should not be normal vector of M at x.

Then the local parameterization involves solving the following augmented system of equations for a given value of η which is a scalar.

$$H(x) \equiv \begin{pmatrix} F(x) \\ e^T(x - x_c) \end{pmatrix} = \begin{pmatrix} 0 \\ \eta \end{pmatrix} \qquad (3.53)$$

The Jacobian

$$DH(x) = \begin{pmatrix} DF(x) \\ e^T \end{pmatrix}$$

is nonsingular in an open neighborhood of $x = x_c$ [13].

This general setting becomes background material for the continuation power flow discussed in section 3.4.   Basically DH(x) can be related to the augmented jacobian J_{aug}. How to choose vector e and the scalar parameter η are discussed in the predictor and corrector tracing process in the same section.

This basic manifold approach can be exploited to identify and trace voltage stability boundaries

Power system equilibrium manifold is defined in this chapter for power system equilibrium tracing.

Saddle node bifurcation related voltage stability margin boundary manifold is defined in Chapter 5 for voltage stability margin boundary tracing.

3.8.4 Equilibrium manifold Tracing of power systems

Simultaneously solving for X and Y will enable us to avoid the assumptions used in power flow. This leads us to the question whether it is possible to solve for X and Y directly and simultaneously from Eqs.3.13 and 3.14 at equilibrium, i.e.,

$$0 = F(X,Y,U,Z) \tag{3.54}$$

$$0 = G(X,Y,U,Z) \tag{3.55}$$

The immediate concern is whether the Newton's method would work with as good convergence as that in the power flow.

As mentioned earlier, the release of slack bus generation is used in power flow so that network losses corresponding to a set of system voltages are not causing convergence trouble during iterations. In the complete description of the system at equilibrium state, this compensation becomes possible without the necessity of removing the slack bus power balance equations. With the description of the system at steady state by 3.54 and 3.55, generation at terminal interface to the network is now a function of system states (see Eqs.3.47). The governor frequency regulation together with the boiler valve control, as described by Eqs.3.35 and 3.34, interacts with the network real power balance constraints, through mechanical power P_{Mi} (Eqs.3.23 and 3.34), to adjust the interface generation P_{gi} so that real power losses are automatically compensated by regulating the system frequency. Similarly, the automatic voltage regulator (described by Eqs.3.14 and 3.16) interacts with the network reactive power balance constraints, through E_{fdi} to adjust Q_{gi} so that reactive power losses are compensated by regulating terminal bus voltage V_i. In regard to PV bus assumption, it is not needed any more since AVR is actually represented.

Based on the above analysis, it is possible to solver for X and Y simultaneously by directly applying Newton's method to Eqs.3.54 and 3.55. Further, in the following section, we will show how we can incorporate this into continuation and apply the resultant simultaneous equilibria tracing technique to voltage collapse identification. Overall solution methodology is given in the sequel.

Eqs.3.54 and 3.55 define the equilibrium manifold of power system. The conventional power flow solution is simply a point on this manifold corresponding to certain condition. It could be thought of as an intersection point of the equilibrium manifold and a cut line (or hyper-plane) defined by system condition. Naturally power system condition is parameterized by control variables U and loading condition Z that present in the power system DAE model.

The equilibrium is the solution of a set of nonlinear equations which are introduced in the previous sections. It could be calculated by Gauss-Sedel method or Newton-Raphson method (or their derivatives). Newton-Raphson type of method is widely used due to its super linear convergence rate. But when load stress on power system is increased, both methods can diverge however close the initial guess is. This is caused by singularity of the total system Jacobian of (Eqs.3.54 and 3.55).

Similar approach to the continuation power flow presented in Section 3.4 can be also applied to trace the total equilibrium as defined by (Eqs. 3.54 and 3.55).

To trace this equilibrium, first we need an initial starting point. Next section provides details related to this initial condition.

3.8.5 Initialization for power system equilibrium tracing

To start power system equilibrium tracing, we need initial conditions that are defined by following variables at all buses $\delta, \omega, E_q', E_d', E_{fd}, V_r, R_f,$ I_d, I_q, V, θ. Solution from power flow provides V, θ at all buses. The remaining values are obtained as shown as follows [6].

The first step in computing the initial conditions is to obtain the generator currents from Eq.3.55:

$$I_{G_i} e^{j\gamma_i} = \frac{P_{G_i} - Q_{G_i}}{V_i e^{-j\theta_i}} \tag{3.56}$$

and the relative machine rotor angels from manipulation of the stator and flux equations

$$\delta_i = \text{angle of } (V_i e^{j\theta_i} + (R_{s_i} + jX_{q_i})I_{G_i} e^{j\gamma_i}) \tag{3.57}$$

With these quantities, the remaining dynamic and algebraic states can be obtained by

$$I_{d_i} + jI_{q_i} = I_{G_i} e^{j(\gamma_i - \delta_i + 90°)} \tag{3.58}$$

$$V_{d_i} + jV_{q_i} = V_i e^{j(\theta_i - \delta_i + 90°)} \tag{3.59}$$

followed by E_{fd} from the stator and flux equation

$$E_{fd_i} = X_{d_i} I_{d_i} + V_{q_i} + R_{s_i} I_{q_i} \tag{3.60}$$

With this field voltage, R_{f_i}, V_{R_i} and V_{ref_i} can be found from the exciter equations as

$$R_{f_i} = \frac{K_{f_i}}{T_{f_i}} E_{fd_i} \tag{3.61}$$

$$V_{r_i} = (K_{e_i} + S_{e_i}(E_{fd_i}))E_{fd_i} \tag{3.62}$$

$$V_{ref_i} = V_i + \frac{V_{r_i}}{K_{a_i}} \tag{3.63}$$

This initial value of E'_{qi} and E'_{di} are then found from the flux equations:

$$E'_{qi} = -(X_{d_i} - X'_{d_i})I_{d_i} + E_{fd_i} \tag{3.64}$$

$$E'_{di} = -(X_{q_i} - X'_{q_i})I_{q_i} \tag{3.65}$$

This completes the computation of all dynamic state initial conditions.

3.8.6 Continuation method with local parametrization

This section extends the application of the continuation method described in Section 3.4 to the power system DAE formulation. The system equilibrium manifold defined by Eqs.3.54 and 3.55 could be traced, according to a scheduled scenario parameterized by λ, from base case up to the point where voltage collapse associated with the saddle node bifurcation occurs.

The same predictor and corrector process described in Section 3.4 can be applied here. Then the tangent vector is solved from

$$\begin{bmatrix} F_X & F_Y & F_\lambda \\ G_X & G_Y & G_\lambda \\ & e_k^T & \end{bmatrix} \begin{bmatrix} dX \\ dY \\ d\lambda \end{bmatrix} = \begin{bmatrix} 0 \\ 0 \\ \pm 1 \end{bmatrix} \qquad (3.66)$$

Once the prediction is made with the tangent vector, the following correction is performed to find the equilibrium point.

$$\begin{bmatrix} F_X & F_Y & F_\lambda \\ G_X & G_Y & G_\lambda \\ & e_k^T & \end{bmatrix} \begin{bmatrix} \Delta X \\ \Delta Y \\ \Delta \lambda \end{bmatrix} = -\begin{bmatrix} F \\ G \\ 0 \end{bmatrix} \qquad (3.67)$$

where $[dX^T, dY^T, d\lambda]^T$ is the tangent vector. e_k is a column unit vector with all the elements equal to zero except for the kth one, which corresponds to the current continuation parameter. Since F_λ and G_λ can not be null vectors at the same time even at the base case ($\lambda = 0$), the singularity of the augmented Jacobian matrix can be easily avoided by appropriately selection of the continuation parameter. To speed up the computation, the same Jacobian can be used in Eqs.3.66 and 3.67.

Since λ is introduced to parameterize the system generation and load level, it increases monotonically to the maximum value. Hence $d\lambda$ is positive before λ reaches its maximum, and negative afterwards. Null $d\lambda$ indicates that the system total Jacobian matrix is singular. This can be clearly seen as follows.

3.8.7 Linerization of power system DAE

When the parameter in Eqs.3.13 and 3.14 is varied, the corresponding state vector X and the eigenvalues of the system matrix evaluated on this path change accordingly.

Linearization of Eqs.3.13 and 3.14 at the equilibrium point with specified U and Z is presented as follows:

$$\begin{bmatrix} \Delta \dot{X} \\ 0 \end{bmatrix} = \begin{bmatrix} F_X & F_Y \\ G_X & G_Y \end{bmatrix} \begin{bmatrix} \Delta X \\ \Delta Y \end{bmatrix} = J_{total} \begin{bmatrix} \Delta X \\ \Delta Y \end{bmatrix} \tag{3.68}$$

Matrices F_X, F_Y, G_X, and G_Y contain first derivatives of F and G with respect to X and Y, evaluated at the equilibrium point.

Note that matrix G_Y is an algebraic Jacobia matrix that contains the power flow Jacobian matrix. In the above equation, if $\det(G_Y)$ does not equal zero

$$\Delta Y = -G_Y^{-1} G_X \Delta X \tag{3.69}$$

Substituing in (3.68) results in

$$\Delta \dot{X} = A_{sys} \Delta X \tag{3.70}$$

$$A_{sys} = F_X - F_Y G_Y^{-1} G_X \tag{3.71}$$

The essential small-disturbance dynamic characteristics of a structure-preserving model are expressed in terms of eigen-properties of the reduced system matrix A_{sys}. This matrix is called dynamic system state matrix.

Eigenvalue analysis of A_{sys}, will give small signal stability information of the current equilibrium point under small disturbances. At voltage collapse, the system loses the ability to supply enough power to a heavily loaded network. At that point, the so-called saddle node bifurcation occurs which is described by the movement of one eigenvalue of A_{sys} on the real axis crossing the origin from the left half complex plane. Eigenvalue computation will help detect this movement, participation factor studies will show how bus voltages participate in this collapse mode, and sensitivity analysis will show the parameter influence on this critical situation [5].

At saddle node bifurcation which leads to voltage collapse, one of the eigenvalue of A_{sys} becomes zero. Equivalently, the determinant of A_{sys} equal zero. From matrix theory, we know that,

$$\det(J_{total}) = \det\begin{bmatrix} F_X & F_Y \\ G_X & G_Y \end{bmatrix} = \det(F_X - F_Y G_Y^{-1} G_X)\det(G_Y) \tag{3.72}$$

$$= \det(A_{sys})\det(G_Y)$$

So if G_Y is nonsingular, the determinant of A_{sys} becomes zero if and only if the determinant of J_{total} is zero. This is the Schur formula. J_{total} is very sparse and thus allow efficient handling using sparse techniques. Therefor detection of the singularity of A_{sys} is equivalent to the detection of the singularity of J_{total}.

3.8.8 Detection of   Saddle Node Bifurcation with System Total Jacobian

Proposition 1:   When G_Y^{-1} exists and $u_X \neq 0$, then the   following equivalent condition is valid

$$A_{sys} u_X = \lambda u_X \tag{3.73}$$

if and only if

$$\begin{bmatrix} F_X - \lambda I & F_Y \\ G_X & G_Y \end{bmatrix}\begin{bmatrix} u_X \\ u_Y \end{bmatrix} = 0 \tag{3.74}$$

where

$$u_Y = -G_Y^{-1} G_X u_X \tag{3.75}$$

We *define the extended eigenvector* $u = \begin{bmatrix} u_X^T & u_Y^T \end{bmatrix}^T$.

Proof:

1. Assume $A_{sys} u_X = \lambda u_X$, i.e.,

$$(F_X - F_Y G_Y^{-1} G_X)u_X = \lambda u_X \tag{3.76}$$

From L.H.S of Eq.3.74

$$\begin{bmatrix} F_X - \lambda I & F_Y \\ G_X & G_Y \end{bmatrix}\begin{bmatrix} u_X \\ u_Y \end{bmatrix} = \begin{bmatrix} (F_X - \lambda I)u_X + F_X u_Y \\ G_X u_X + G_Y u_Y \end{bmatrix} \qquad (3.77)$$

$$= \begin{bmatrix} (F_X - \lambda I)u_X - F_X G_Y^{-1} G_X u_X \\ G_X u_X - G_Y G_Y^{-1} G_X u_X \end{bmatrix} = \begin{bmatrix} (F_X - F_X G_Y^{-1} G_X)u_X - \lambda u_X \\ 0 \end{bmatrix} = 0$$

Substitution of $u_Y = -G_Y^{-1} G_X u_X$ in the above equation verifies Eq.3.74.
Or

2. Assume

$$\begin{bmatrix} F_X - \lambda I & F_Y \\ G_X & G_Y \end{bmatrix}\begin{bmatrix} u_X \\ u_Y \end{bmatrix} = 0 \qquad (3.78)$$

$$\begin{bmatrix} (F_X - \lambda I)u_X + F_X u_Y \\ G_X u_X + G_Y u_Y \end{bmatrix} = 0 \qquad (3.79)$$

From the second item in Eq.3.79, $u_Y = -G_Y^{-1} G_X u_X$. Substitute this into the first item

$$(F_X - \lambda I)u_X - F_X G_Y^{-1} G_X u_X = 0 \qquad (3.80)$$

After rearrangement, based on the definition of A_{sys}

$$A_{sys} u_X = \lambda u_X \qquad (3.81)$$

is obtained. This concludes the proof for proposition 1.

□

From Eq.3.74, the total Jacobian matrix

$$A_{total} = \begin{bmatrix} F_X & F_Y \\ G_X & G_Y \end{bmatrix}$$

can be used to detect either Saddle node or Hopf bifurcation.

3.8.8.1 Detection of saddle-node bifurcation

From proposition 1, the condition

$$\begin{bmatrix} F_X & F_Y \\ G_X & G_Y \end{bmatrix}\begin{bmatrix} u_X \\ u_Y \end{bmatrix} = 0 \qquad (3.82)$$

can be utilized to detect Saddle node bifurcation, that is, to detect the singularity of the total Jacobian matrix.

During the direct equilibrium tracing, the saddle node bifurcation point can be readily identified by utilizing a cut function, without computing eigenvalues [13,15].

A cut function for Saddle node related fold bifurcation can be implicitly defined as $\gamma_{SNB}(x)$ in the following equation:

$$\begin{bmatrix} F_X & F_Y & e_k \\ G_X & G_Y & \\ e_j^T & & 0 \end{bmatrix} \begin{bmatrix} u_X^0 \\ u_Y^0 \\ \gamma_{SNB} \end{bmatrix} + \begin{bmatrix} 0 \\ 0 \\ 1 \end{bmatrix} = 0 \tag{3.83}$$

where we denote $u^0 = [u_X^0 \quad u_Y^0]^T$. Or equivalently,

$$\begin{bmatrix} F_X & F_Y & e_k \\ G_X & G_Y & \\ e_j^T & & 0 \end{bmatrix}^T \begin{bmatrix} v_X^0 \\ v_Y^0 \\ \gamma_{SNB} \end{bmatrix} + \begin{bmatrix} 0 \\ 0 \\ 1 \end{bmatrix} = 0 \tag{3.84}$$

where we denote $v^0 = [v_X^0 \quad v_Y^0]^T$.

At the fold point, the cut set condition is satisfied, that is $\gamma_{SNB}(X,Y,\alpha,\beta) = 0$.

At each continuation step, γ_{SNB} is checked. If γ_{SNB} changes sign, Saddle node bifurcation has just been passed. This γ_{SNB} is nothing but $d\lambda$ [16].

If null $d\lambda$ is detected at some step, then Eq.3.36 reduces to

$$\begin{bmatrix} F_X & F_Y \\ G_X & G_Y \end{bmatrix} \begin{bmatrix} dX \\ dY \end{bmatrix} \approx [J_{total}] \begin{bmatrix} dX \\ dY \end{bmatrix} = \begin{bmatrix} 0 \\ 0 \end{bmatrix} \tag{3.85}$$

Since

$$[e_k^T] \begin{bmatrix} dX \\ dY \\ 0 \end{bmatrix} = \begin{bmatrix} 0 \\ 0 \\ \pm 1 \end{bmatrix} \tag{3.86}$$

So one of the components of dX or dY is ± 1, not a null vector, Eq.3.85 hence implies the total system Jacobian J_{total} singular. As mentioned before, from Eq.3.72, the singularity of J_{total} coincides with the singularity of A_{sys} if G_Y is nonsingular.

The singularity of A_{sys} implies it has a null eigenvalue at the current step. Therefore null $d\lambda$ exactly signifies a saddle node bifurcation. Thus it can readily identify the saddle node bifurcation point by equivalently detecting null $d\lambda$ during the direct equilibrium tracing, without formation of A_{sys} and computing its eigenvalues.

However, when system limits are considered, sometimes we could not capture the null $d\lambda$ point even using a very small step length. It most probably means an immediate voltage collapse encountered due to some generators hitting their limits [17]. On the other hand, in order to investigate the voltage collapse mechanisms or to develop an effective control strategy against voltage collapse, the critical eigenvalue responsible for the voltage collapse may be needed.

In general, there is no simple way to capture the critical eigenvalue at an immediate voltage collapse point. However, this critical eigenvalue can readily be detected via simultaneous equilibrium tracing. That is, we can use the general tracing scheme illustrated in Fig.3.24 to locate the saddle node bifurcation point where the critical eigenvalue crosses the origin on the complex plane.

First, we use a relatively large step size to trace the system equilibrium diagram BC_{im} until the negative $d\lambda$ is detected at point C_{im}. Then we should change the tracing direction and continue the process with a smaller step size up to the saddle node bifurcation point C_{snb} where null $d\lambda$ could be easily detected. If the traced equilibrium diagram is the same as depicted in Fig.3.24, we can conclude that the point C_{im} is the system immediate voltage collapse point. Otherwise, in case null $d\lambda$ is detected but the saddle node bifurcation point C_{snb} is sitting on the BC_{im} diagram, it means that the voltage collapse results from the saddle node bifurcation rather than the system limits. Note that the tracing process always stops at the saddle node bifurcation point. The solid curve with arrows indicates the tracing path and direction.

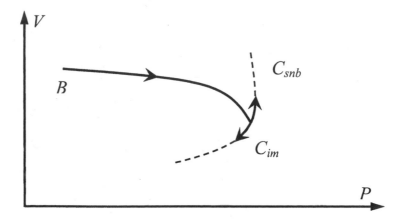

Fig. 3.24 Illustration of direct equilibrium tracing process

3.8.9 Limits implementation

It is very important to reasonably represent the system limits when studying voltage stability. In fact, voltage collapse occurs more than often as a consequence of limited local reactive power supply. When the system loses the ability to further meet the load demand in a heavily stressed network, the cascaded hitting of limits usually leads to system collapse. There are basically two types of limits to be considered. One is the governor limit, and the other is the AVR output limit. For voltage stability, the latter usually plays a more important role.

3.8.9.1 Governor limits

The governor limits are implemented by regulating the real power generation/load settings. Those generators which hit P_{gsi}^{max} will then be forced to stay at maximum, and no longer allowed to further pick up the system load increase.

3.8.9.2 AVR limits

The automatic voltage regulator (AVR) controls the terminal voltage of the synchronous machine. It indirectly controls the reactive power output by regulating the AVR output voltage V_r. In the new formulation, we are able to directly implement the limits which are usually given to restrict the output of the voltage regulator. Forcing the AVR output voltage at a particular value will directly control the rotor current to stay below limits and indirectly control the reactive generation. This can be shown as follows. At an equilibrium state, the AVR output voltage is related to the synchronous machine rotor current as

$$V_{ri} = (K_{ei} + S_{ei})E_{fdi} \tag{3.87}$$
$$= (K_{ei} + S_{ei})E_{qi}$$
$$= (K_{ei} + S_{ei})X_{adi}I_{fdi}$$

where E_{qi} is the generator's internal induced quadrature axis voltage [18]. So if we ignore the saturation effect, the rotor current is proportional to V_{ri}, which verifies the first half of the above statement. A machine's reactive power output can be written as:

$$Q_{Gi} = \frac{V_i E_{qi}}{X_{di}}\cos(\delta_i - \theta_i) - V_i^2 (\frac{\cos^2(\delta_i - \theta_i)}{X_{di}} + \frac{\sin^2(\delta_i - \theta_i)}{X_{qi}}) \tag{3.88}$$

When V_{ri} is fixed at a certain value, the reactive power will then be limited indirectly, at lease not increase exponentially when approaching voltage collapse. This shows the second half of the previous statement.

Once the AVR of a generator hits the limit, it loses the ability to adjust V_{ri} and thus Q_{Gi} to meet the load increase. The AVR has to be set so that V_{ri} stays at the limiting value. Referring to Eq.3.32, the dynamic differential equation will be dropped and will not be included for stability analysis. This is obvious if one recalls the definition of stability from control theory. That is, the limited dynamic state will stay as a constant, and it no longer participates in the dynamic response of the system. If we solve the remaining equations which provide the DAE description of the system with the same control inputs, we may not be able to find a solution. This is because, when the system load further increases, in order to continuously keep V_{ri} at the limiting value, the corresponding excitation reference voltage V_{refi} may have to be reduced. The decrease of the exciter reference voltage reflects the inability of the generator to keep pace with the load increase. In the conventional two-step based equilibria tracing approach, this would require a new power flow solution with a different set of generation and/or voltage specifications for the PV buses. After this, the X variables are then calculated and the control inputs including the exciter reference voltage will then be updated to a new smaller value in this case. As mentioned in the first section of this chapter, this causes the problem of inconsistent description of the generators. In the new formulation, when some new limits are hit, this update of control settings can be done automatically during continuation. To do so, we include the following equation, which is nothing but the right hand side of Eq.3.32 with V_{ri} at its maximum.

$$0 = \frac{1}{T_{ai}}(-V_{ri}^{max} + K_{ai}(V_{refi} - V_i - R_{fi})) \cong f_i^{AVR+} \tag{3.89}$$

If a new limit is found to be violated at the end of the current correction, the following Jacobian will then be used in the immediate correction to update the input exciter reference voltage.

$$- \begin{bmatrix} \overline{F}_{\overline{X}} & \overline{F}_Y & 0 & \overline{F}_\alpha \\ G_{\overline{X}_i} & G_Y & 0 & G_\alpha \\ f_{i\overline{X}}^{AVR+} & f_{iY}^{AVR+} & f_{iV_{ref}}^{AVR+} & 0 \\ & e_k^T & & \end{bmatrix} \begin{bmatrix} \Delta \overline{X} \\ \Delta Y \\ \Delta V_{refi} \\ \Delta \alpha \end{bmatrix} = \begin{bmatrix} \overline{F} \\ G \\ f_i^{AVR+} \\ 0 \end{bmatrix} \tag{3.90}$$

where $\overline{F} = \{F\} - \{f_i^{AVR}\}$ and $\overline{X} = \{X\} - \{V_{ri}\}$ and $f_{iV_{ref}}^{AVR+} \cong \partial f_i^{AVR+}$

$/ \partial V_{refi}$. After this, if no new limits are violated, the following equation will then be used for subsequent correctors:

$$- \begin{bmatrix} \overline{F}_{\overline{X}} & \overline{F}_Y & 0 & \overline{F}_\lambda \\ G_{\overline{X}} & G_Y & 0 & G_\lambda \\ f_{i\overline{X}}^{AVR+} & f_{iY}^{AVR+} & 10^{15} & 0 \\ & e_k^T & & \end{bmatrix} \begin{bmatrix} \Delta \overline{X} \\ \Delta Y \\ \Delta V_{ri} \\ \Delta \lambda \end{bmatrix} = \begin{bmatrix} \overline{F} \\ G \\ 0 \\ 0 \end{bmatrix} \tag{3.91}$$

Once the limit is hit, the predictor equation from then on is changed to

$$-\begin{bmatrix} \overline{F}_{\overline{X}} & \overline{F}_Y & 0 & \overline{F}_\lambda \\ G_{\overline{X}} & G_Y & 0 & G_\lambda \\ f_{i\overline{X}}^{AVR+} & f_{iY}^{AVR+} & 10^{15} & 0 \\ & e_k^T & & \end{bmatrix}\begin{bmatrix} d\overline{X} \\ dY \\ dV_{ri} \\ d\lambda \end{bmatrix} = \begin{bmatrix} 0 \\ 0 \\ 0 \\ \pm 1 \end{bmatrix} \tag{3.92}$$

The large number is used to keep the size of the matrix unchanged which provides programming ease. And by using this Jacobian, we observe that neither the AVR output voltage nor the input exciter reference voltage is updated during the prediction process. This makes sure that we get the tangent of the equilibrium curve corresponding to the current input settings while satisfying the limits already encountered. The above analysis is illustrated in Fig.3.25.

When $d\lambda$ is zero, from Eq.3.92:

$$det\begin{pmatrix} \overline{F}_{\overline{X}} & \overline{F}_Y & 0 \\ G_{\overline{X}} & G_Y & 0 \\ f_{\overline{X}}^{AVR+} & f_Y^{AVR+} & 10^{15}I \end{pmatrix} = 0 \tag{3.93}$$

And we have

$$det\begin{pmatrix} \overline{F}_{\overline{X}} & \overline{F}_Y & 0 \\ G_{\overline{X}} & G_Y & 0 \\ f_{\overline{X}}^{AVR+} & f_Y^{AVR+} & 10^{15}I \end{pmatrix} = det\begin{pmatrix} \overline{F}_{\overline{X}} & \overline{F}_Y \\ G_{\overline{X}} & G_Y \end{pmatrix} det(10^{15}I) \tag{3.94}$$

Thus we observe that $d\lambda = 0$ again signifies saddle node bifurcation of the DAE model.

The above derivation [19] provides the validity of using the iterative continuation of Jacobian (Eq.3.92) in simultaneous equilibria tracing to identify voltage collapse, both before and after hitting AVR limits. In Chapter 4, we will see that the continuation Jacobian can also be used for studying the sensitivity of the saddle node bifurcation of the DAE model.

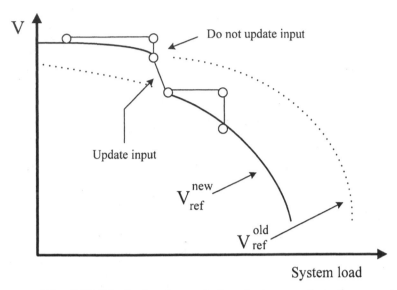

Fig. 3.25 Limits implementation during continuation

3.9 Numerical examples for EQTP

EQTP Scenario Description:

In the EQTP simulation, the scenario is similar to scenario 1 in CPF. Loads at the same 8 buses are increased, while the increased load is picked up by the same 9 generators. And both load and generator are increased proportionally by their initial load and generation levels.

For this scenario the variations of load bus voltages, generator real powers, and reactive powers for changing load are shown in Figs. 3.26, 3.27, and 3.28 respectively.

As explained in the previous sections, the automatic voltage regulator regulates the generator terminal voltage and its reactive power output of the network. The speed governor adjusts the real power generation and frequency to meet load increase. Because all these devices are modeled in detail, we are able to observe how the synchronous machines interact with the network, both before and after hitting the limits. The inability of indefinitely supplying power through the network to the load centers, as consequence of control system or machine capacity limitations and network loadability restrictions, will ultimately lead to system voltage collapse.

For all the generators which hit their AVR output voltage limits, the terminal voltage, AVR output voltage, reactive power generation, and exciter reference voltage have similar response profiles. Therefore we take the generator at bus 30 bus as the example for the explanation.

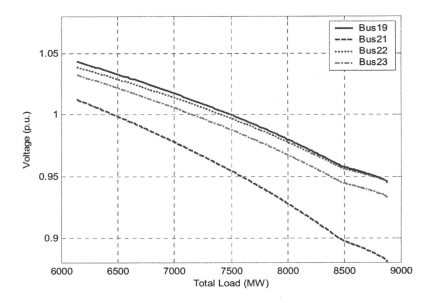

Fig. 3.26 V vs. P$_{total}$

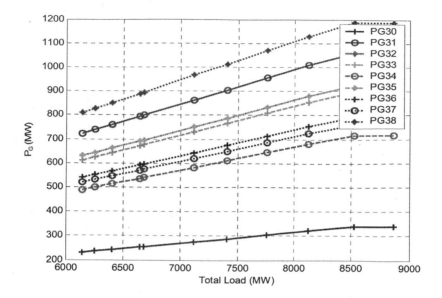

Fig. 3.27 P_G vs. P_{total}

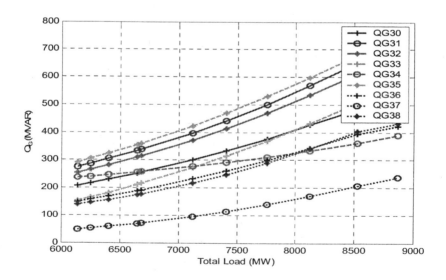

Fig. 3.28 Q_G vs. P_{total}

Fig 3.29 shows that, before hitting its AVR output limit, the voltage regulator can maintain a fairly high and steady terminal voltage. When the system total load exceeds 8845MW, AVR output voltage as shown in Fig. 3.30 hits the maximum value and the terminal voltage experiences a noticeable drop.

Fig. 3.31 shows the profile of reactive power generation at bus 32. A sudden slowing down of the increase in the reactive power generation occurs when the AVR output limit is hit. From this point on, fixing the AVR output voltage makes the terminal reactive power generation to decrease slightly.

Figs. 3.30 and 3.32 are the AVR output and exciter reference voltages of the generator at bus 32.

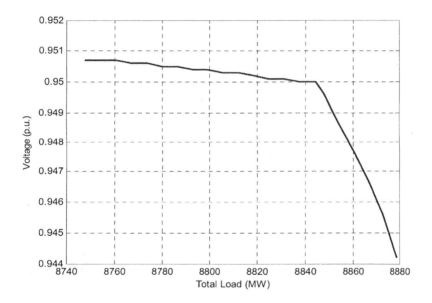

Fig. 3.29 Voltage at Generator Bus 32

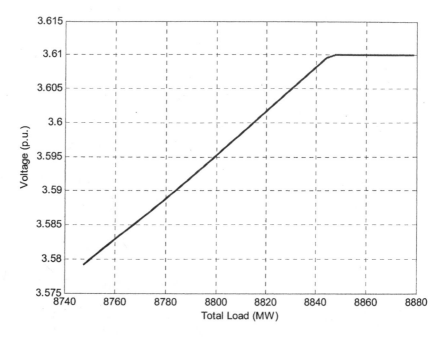

Fig. 3.30 AVR Output Voltage Vr at Bus 32

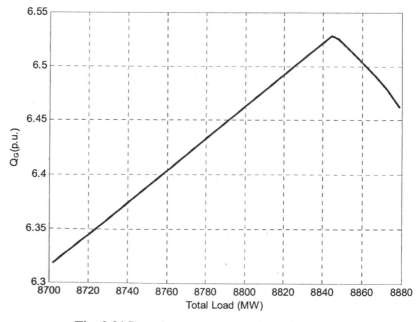

Fig. 3.31 Reactive power generation at bus 32

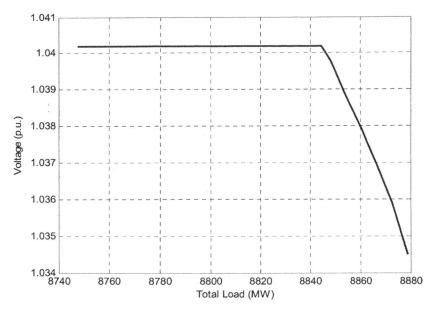

Fig. 3.32 Exciter reference voltage at bus 32

The governor associated with generator at bus 30 is the first to reach its limit when the system loading level is 8523 MW. Fig. 3.33 shows the governor setting value. At a system loading level of 8523 MW, when most of the governors hit their limits, the system frequency experiences sag as shown in Fig. 3.34.

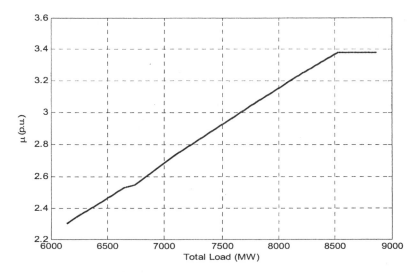

Fig. 3.33 Governor Response of Generator at Bus 30

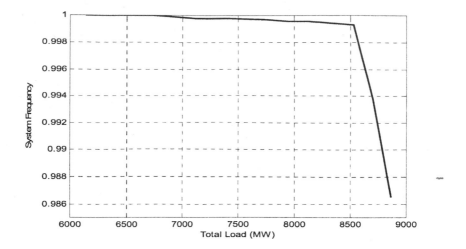

Fig. 3.34 System Frequency Response

As mentioned before various versions of continuation power flow methods are proposed in the literature [22-26]. These methods depend on the type of parameterization strategies. Continuation based approaches can also be used for critical eigenvalue tracing. References [12, 29, 30] discuss the application of continuation based approaches for eigenvalue tracing.

References

[1]   Allgower, E. L. and Gearg, K., *Numerical Continuation Methods.*
      Springer-Verlag, 1990

[2]   Seydel, R., *From equilibrium to chaos,* Elsevier Science, 1988

[3]   Rheinboldt, W.C., "Solutions fields of nonlinear equations and con-
      tinuation methods, " *SIAM J. Num. Anal.*, Vol. 17, 1980, pp. 221-237

[4a] Christy, C.D., *Analysis of steady state voltage stability in large scale
      power systems*, M.S. Thesis, Iowa State University, Ames, IA, 1990

[4b] Ajjarapu, V. and Christy, Colin, "The Continuation Power Flow: A
      Tool to Study Steady State Voltage Stability," *IEEE Transactions on
      Power Systems*, Vol. 7, No. 1, pp. 416-423, Feb. 1992

[5]   Kundur, P., *Power system stability and control,* McGraw Hill, 1994

[6]   Sauer, P. W., Pai, M. A., "Power system steady-state stability and the
      load flow jacobian, " *IEEE Trans. on power systems*, 1990, 1374-1383

[7]   Arrillaga, J., Smith, B., *AC-DC power systems analysis,* The Institu-
      tion of Electrical Engineers, London, UK, 1998

[8]   Arabi, S. and Kundur, P.,"A versatile FACTS device model for power
      flow and stability simulations " IEEE Transactions on Power Systems ,
      Volume 11, Issue 4, Nov. 1996 Page(s):1944 – 1950

[9]   Iowa State University's Web Based Voltage Stability Search Engine
      http://design-2.ece.iastate.edu/biblio/

[10] Guckenheimer, J., Holmes, P.J., *Nonlinear oscillations, dynamical sys-
      tems, and bifurcation of vector fields*, New York: Springer-Velag,
      1983

[11] Illic, M., Zaborsky, J., *Dynamics and control of large electric power
      systems*, John Wiley & Sons, New York, 2000

[12] Kim, K., Schattler, H., Venkatasubramanian, V., Zaborsky, J., Hirsch,
      P., "Mehods for calculating oscillations in large power systems," *IEEE
      Trans. on power systems*, Vol.12, 1997, pp.1639-1648

[13] Rheinboldt, W.C., *Numerical analysis of parameterized nonlinear
      equations*, New York: John Wiley & Sons, 1986

[14] Souza, De A.C.Z., Canizares, C.A., Quintana, V.H., "New techniques
      to speed up voltage collapse computations using tangent vectors,"
      IEEE Trans. on power systems, Vol.12, 1997, 1380-1387

[15] Dai, R., Rheinboldt, W.C., "On the computation of manifolds of fold-
      points for parameter-dependent problems," *SIAM J. Numer. Anal.*,
      Vol. 27, no.2, 1990, pp. 437-446.

[16] Zhou, Y., Ajjarapu, V., "A fast algorithm for identification and tracing
      of voltage and oscillatory stability margin boundaries," *Proceedings of
      the IEEE*, Vol. 93, no. 5, 2005, pp.934 – 946

[17] Dobson, I., Lu, L., "Voltage collapse precipitated by the immediate change in stability when generator reactive power limits are encountered," *IEEE Trans. on Circuits and Syst.*, Vol.39:9, pp.762-766, 1992

[18] Anderson, P.M., Fouad, A.A., *Power System Stability and Control*, New York: IEEE PES Press, 1994

[19] Long, B., *The detection of dynamic voltage collapse and transfer margin estimation*, M.S. Thesis, Iowa State University, Ames, IA, 1997

[20] Hill, D. J., Nonlinear dynamic load models with recovery for voltage stability studies, *IEEE Trans. on Power Syst.*, vol.8:1, pp.166-172, 1993

[21] Van Cutsem, T., Vournas, C., Voltage stability of electric power systems, Kluwer, 1998

[22] Iba, K., Suzuki, H., Egawa, M., Watanabe, T., *Calculation of critical loading condition with nose curve using homotopy continuation method*, IEEE Transactions on Power Systems, Volume 6, Issue 2, pp.584 – 593, May 1991

[23] Canizares, C. and A., Alvarado, "Point of collapse and continuation methods for large AC/DC power systems," *IEEE Transactions on Power Systems*, vol.8, no.1, February 1993, pp.1-8

[24] Canizares, C. A., Alvarado, F. L., Demarco, C. L., Dobson, I. and Long, W. F., "Point of collapse methods applied to AC/DC power systems," *IEEE Trans. on Power Systems*, vol.7, no.2, pp. 673–683, 1992

[25] Chiang, H.D., Flueck, A.J., Shah, K.S., and Balu, N., "CPFLOW: a practical tool for tracing power system steady-state stationary behavior due to load and generation variations," *IEEE Trans. on power systems*, PWRS-10, May 1995, pp.623-634

[26] Flueck, A.J., Dondeti, J.R., "A new continuation power flow tool for investigating the nonlinear effects of transmission branch parameter variations," *IEEE Trans. on power systems*, vol.15:1, pp.223-227, 2000

[27] Denis Lee Hau Aik and Anderson, G " Quasi-static stability of HVDC systems considering dynamic effects of synchronous machines and excitation voltage control ," IEEE transactions on Power Delivery , Volume 21, Issue 3, July 2006 pp:1501 – 1514

[28] Canizares, C.A. and Faur, Z.T " Analysis of SVC and TCSC controllers in voltage collapse" IEEE Transactions on Power Systems , Volume 14, Issue 1, Feb. 1999 Page(s):158 – 165

[29] Wen, X., Ajjarapu, V., "Application of a Novel Eigenvalue Trajectory Tracing Method to Identify Both Oscillatory Stability Margin and Damping Margin," IEEE Transactions on Power Systems, Volume 21, May 2006, pp. 817-824

[30] Yang, D. and Ajjarapu, V., "Critical Eigenvalues Tracing for Power System Analysis via Continuation of Invariant Subspaces and Projected Arnoldi Method," to appear in IEEE Trans. on Power Systems

4 Sensitivity Analysis for Voltage Stability

4.1 Introduction

In power system voltage stability analysis, it is not enough to merely obtain the critical point. It is important to know how this critical point is affected by changing system conditions. One should get the information related to parameters and controls that may influence the system stability.

The intention is to get a measure of the margin between the current point of operation and the point where the system becomes unstable, thereby providing early warning of a potentially critical condition. Sensitivity analysis becomes a major part in defining this measure. The attributes of this measure depends on various factors. This measure can be in the form of an index. The indices in use are very different and it is convenient to classify the indices into two main classes (given-state based and large deviation based indices) as suggested by [1]. Recent IEEE special publication on voltage stability [2] has an excellent and exhaustive description of various indices.

The DAE model together with the corresponding system state matrix A_{sys} provides true dynamic stability information. A static power flow based analysis does not give true stability information. Keeping this in mind, one should be clear that, for instance, the minimum eigenvalue can be that of A_{sys} and J_{LF} (power flow Jacobian).

4.2 Given State Based Indices [1]

These indices only use the information available at the current operating point. The operating point could be simulated for a desired power transfer condition. From this information, the system characteristic is calculated and system operation is classified.

- Reactive power reserve

Automatically activated reactive power reserve at effective locations can serve as a simple, yet sensitive, voltage security index. And in addition to being a given state index, it can also serve as a large deviation index (MVAR distance to voltage collapse), with the assumption that instability occurs when the field current of a key generator reaches its limit or when a SVC reaches its boost limit.

- Voltage drop

These indices are based on the principle that the voltage drops as the system is loaded. However this is sometimes masked by the effect of reactive power compensation devices and off-nominal tap setting of transformers.

- MW/MVAR losses

The losses increase exponentially when a system approaches voltage collapse. The application of these losses used as indicators of voltage instability has been given in the literature.

- Incremental values

These indices give information about the system state in the close vicinity of an operating point. Incremental values can provide a quantitative insight of weakness of a node. $\Delta Q / \Delta V$, for example, is sometimes used for assessing areas prone to voltage collapse.

- Incremental steady state margin

This is an indicator calculated from a determinant of a special formulation of the system power flow Jacobian. After normalization, the maximum index value will be 1.0 and will reach 0.0 at critical load conditions. The earliest form of this index was proposed by Venikov [3].

- Minimum singular value or minimum eigenvalue

Singular values have been employed in power systems because of the useful orthogonal decomposition of the Jacobian matrices. The singular value decomposition is typically used to determine the rank of a matrix, which is equal to the number of non-zero singular values of the matrix.

Consider a $n \times n$ real matrix A. The singular value decomposition (*SVD*) of A is written [4]

$$A = QSP^T = \sum_{i=1}^{n} s_i q_i p_i^t$$

where S is an $n \times n$ diagonal matrix and Q and P are $n \times n$ orthonormal matrices. The diagonal elements of S are called singular values of A. The columns $q_1, q_2, \cdots, q_n$ of Q are called the right singular vectors. The columns $p_1, p_2, \cdots, p_n$ of P are called the left singular vectors. By appropriate choice of Q and P, singular values can be arranged such that $s_1 \geq s_2 \cdots \geq s_n \geq 0$.

For a real symmetric matrix A, the individual singular values are equal to the square root of the individual eigenvalues of $A^T A$ or AA^T [5]. Thus, for a real symmetric matrix, the absolute values of eigenvalues are equal to the singular values. Additionally, the smallest singular value of A is the 2-norm distance of A to the set of all rank-deficient matrices [5]. If the minimum singular values is zero (i.e., $s_n = 0$), then the matrix A is singular.

Hence, its application to static voltage collapse analysis focuses on monitoring the smallest singular value up to the point where it becomes zero. Therefore it has been proposed as an index measuring stability [6, 7].

Thus, in voltage stability studies, the minimum singular value of the Jacobian becoming zero corresponds to the critical mode of the system. In [6], the authors calculated the minimum singular value and the two corresponding (left and right) singular vectors of the power flow Jacobian. They defined a voltage stability index as the minimum singular value of the power flow Jacobian, which indicates the distance between the studied operating point and the steady-state voltage stability limit. Similarly, the minimum eigenvalue could also be used as an index because it also becomes zero at the same time as the minimum singular value does.

When given state based indices are plotted against system load, most of their trajectories assume an exponential curvature. This makes it difficult to effectively predict voltage collapse using these indices [8].

4.3 Large Deviation Based Indices [1]

Large deviation based indices account for nonlinearities caused by larger disturbances or load increases. These indices are normally more computationally demanding than the given state indices, but are more reliable. The margin is usually given in terms of the maximum increase in MW or MVAR load, and can either be based on a smooth increase in load from the normal operating conditions, or the load increase can be combines with contingencies in the system.

Methods based on large deviation indices in principle apply the same measure. However, the approach for calculation is very different. Some main classes are:

• Repeated power flows [1]

• Continuation methods [9, 10,11, 12]

• Optimization-based methods [13]

• Direct methods or Point-of-collapse methods [14, 15]

• Closest distance to maximum transfer boundary [16]

• Energy function methods [17]

4.4 Stability Studies via Sensitivity Analysis

As introduced in chapter 3, the dynamic properties of the power system are characterized by the eigenproperties of the system state matrix A_{sys}. In practical situations, obtaining stability results is only part of the work. It is important to identify the key factors which affect stability, either beneficially or detrimentally. These factors can be described by the parameter influence on system performance and stability. The parameters can be operational or non-operational. A common approach in doing sensitivity analysis is to define a stability index and then study how the different parameters affect this index. By using sensitivity techniques, useful information about the relationships between state, control, and dependent variables can established. These sensitivity signals are valid in the vicinity of the point of linearization. Sometimes the sensitivity might not be directly defined with respect to a certain stability index, and is therefore referred to as parametric sensitivity. Since system performance degradation often leads to loss of stability, parametric sensitivity is also used in sensitivity-based stability analysis. At a normal operating state, sensitivity analysis provides information about how different parameters influence stability. Certain control measures can be designed in order to prevent the system from instability. Should the system be in an emergency state under disturbances, effective controls must be applied to pull the system back to a normal state. Sensitivity analysis is well suited for evaluating the effectiveness of the controls.

4.4.1 Identification of critical elements

Identifying critical elements involves locating the key components in a power system (buses, branches, or generators) that are critical to maintain voltage stability. In other words, one should find the weak areas in the system. Different authors present different approaches for finding such areas.

Near a given equilibrium solution (X_0, Y_0) of the structure preserving power system model as given in chapter 3 (Eqs.3.54 and 3.55), the derivatives $\partial X / \partial P$ and $\partial Y / \partial P$ at P_0 give a natural measure of the sensitivity of the solution. Here, P is a vector which includes all parameters explicitly appearing in F and G. From these derivatives, sensitivities of the dependent variables can be easily found. For instance, bus voltage sensitivity with respect to system load, can all be computed from $\partial X / \partial P$ and $\partial Y / \partial P$. From such sensitivities, a proper direction for adjusting the system control variables can be found.

4.4.2 Eigenvalue sensitivity

As explained in Chapter 3, eigenvalue analysis gives information about small signal stability of the current operating point. Therefore the sensitivity of the critical eigenvalue(s) with respect to system parameters is often needed to design coordinated controls to prevent instability. Suppose λ_i is the critical eigenvalue of interest, its sensitivity with respect to any parameter p is [18]:

$$\frac{\partial \lambda_i}{\partial p} = \frac{v_i^T \frac{\partial A_{sys}}{\partial p} u_i}{v_i^T u_i} \tag{4.1}$$

where u_i and v_i are the right and left eigenvectors of A_{sys} corresponding to λ_i respectively. Eigenvalue sensitivity can be applied to any eigenvalue of critical interest, therefore oscillatory as well as collapse type instability can all be addressed by this approach. For voltage collapse analysis, one can apply this to the minimum zero crossing eigenvalue λ_{min}.

4.4.3 Modal analysis

Proposed by Gao et al. [19], modal analysis involves calculation of eigenvalues and eigenvectors of the power flow Jacobian. With a steady-state power system model, the authors computed a specified number of eigenvalues and the corresponding eigenvectors of the (reduced) Jacobian. Assume ξ_i and η_i are, respectively the right and left eigenvectors of the Jacobian corresponding to the eigenvalue λ_i. Then the ith modal reactive power variation is

$$\Delta Q_{m_i} = K_i \xi_i$$

where $K_i^2 \sum_{j=1}^{n} \xi_{ji}^2 = 1$, and the corresponding i^{th} modal voltage variation is

$$\Delta V_{m_i} = \frac{1}{\lambda_i} \Delta Q_{m_i}$$

If ΔV_{m_i} is known, $\Delta \xi_{m_i}$ can be calculated from the power flow equations. Different participations are defined as follows.

- Bus participations: Participation of bus k to mode i is
$$P_{ki} = \delta_{ki} \eta_{ki}$$
where δ_{ki} is the kth element of the ith column right eigenvector and η_{ki} is the kth element of the ith row left eigenvector.

- Branch participations: The participation of branch lj to mode i is

$$P_{lji} = \frac{\Delta Q_{lji}}{\Delta Q_{l\,max\,i}}$$

where $\Delta Q_{l\,max\,i} = \max(\Delta Q_{lji})$ and ΔQ_{lji} is the linearized reactive loss variation across branch lj.

- Generator participations: The participation of generator gk to mode i is

$$P_{gki} = \frac{\Delta Q_{gki}}{\Delta Q_{g\,\max i}}$$

where $\Delta Q_{g\,\max i} = \max(\Delta Q_{gki})$ and ΔQ_{gki} is the linearized reactive power output variation at generator gk.

In the three foregoing participations, the suffix i indicates a particular mode, i, corresponding to the eigenvalue λ_i. A component with higher participation indicates that this component's contribution to this mode is large. Reference [19] used these participations to identify the buses, branches, and generators contributing to a particular mode, for both a base and a critical case.

The next section explains how the same participation information corresponding to critical mode can be obtained and how a voltage stability index can be derived from the tangent vector of the continuation power flow.

4.4.4 Sensitivity analysis via CPF

In the continuation process described in Chapter 3, the tangent vector is useful because it describes the direction of the solution path at a corrected solution point. A step in the tangent direction is used to estimate the next solution. But if we examine the tangent vector elements as differential changes in bus voltage angles ($d\delta_i$) and magnitudes (dV_i) in response to a differential change in load connectivity ($d\lambda$), the potential for meaningful sensitivity analysis becomes evident. The next examples demonstrate how the tangent vector elements change for different load levels. The numerical results that are presented in this chapter are from the New England 39-bus system. The data and one line digram for this system are given in Appendix A.

Figs.4.1 and 4.2 illustrate the tangent vector elements versus the element number for two cases with different load levels, i.e., base case (light load) and critical case (heavy load). It should be noted that the first half of the graphs (first 38 elements for 39-bus system) corresponds to voltage angle terms and that the next half (elements 39 to 67), corresponds to voltage magnitude terms. If we consider the first half of the graph (i.e., up to the elements that correspond to voltage angle), the voltage angle terms are dominant for the light load condition than for the heavy load condition, Whereas the second half of the graph tells us that the voltage magnitudes are dominant for the heavy load condition than for the light load condition. These results are consistent with the conclusions reached by Lof et al. [6] by performing singular value decomposition of the power flow Jacobian.

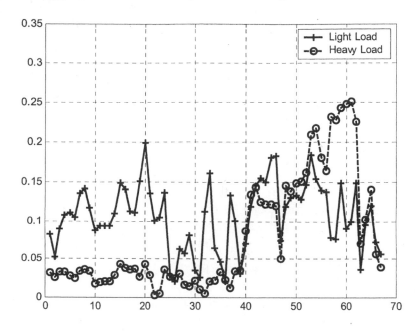

Fig. 4.1. Tangent vector elements for the two cases with different load levels – 39-bus system

4.4.5 Tangent vector, right eigenvector, and right singular vector of J

Examining Eq.3.7 (chapter 3)   from which the tangent vector is calculated, it can be shown that the tangent vector is the right eigenvector of the Jacobian corresponding to zero eigenvalue at the critical point. Additionally, the right eigenvector is equal to the right singular vector because the Jacobian is real and (almost) symmetric. Thus, at the critical point, the tangent vector is equal to the right eigenvector corresponding to the minimum eigenvalue and the right singular vector corresponding to the minimum singular value. This equivalence is evident from Fig.4.2, in which the tangent vector, the right eigenvector, and the right singular vector of the Jacobian matrix near the critical point are plotted on the same graph for the 39 bus system. This information from the tangent vector can be used to identify buses, branches, and generators that are critical to maintain voltage stability. The next section shows how a voltage stability index can be derived from the tangent vector.

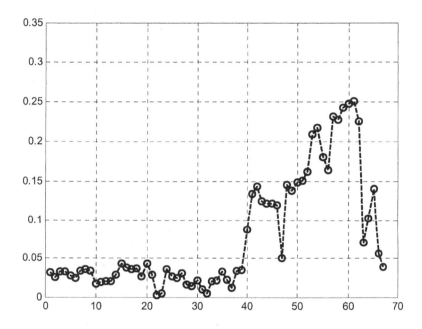

Fig. 4.2 Elements of tangent vector, right eigenvector, and right singular vectors near the critical point – 39-bus system

4.4.6 Voltage stability index from the tangent vector

Voltage stability index using tangent vector can be derived [9]. For this, we have to first find the weakest bus with respect to voltage stability. This is same as finding the bus with the greatest dV_i/dP_{total} value. Here dP_{total} is the differential change in active load for the whole system and is given by

$$dP_{total} = \sum_n dP_{L_i} = [S_{\Delta BASE} \sum_n K_{L_i} \cos(\psi_i)]d\lambda = Cd\lambda \qquad (4.2)$$

The weakest bus, j, is

$$|\frac{dV_j}{Cd\lambda}| = \max[|\frac{dV_1}{Cd\lambda}|,|\frac{dV_2}{Cd\lambda}|,\cdots,|\frac{dV_n}{Cd\lambda}|]$$

where j reaches its steady-state voltage stability limit, $d\lambda$ approaches zero, the ratio $dV_j/Cd\lambda$ becomes infinite or equivalently the ratio $Cd\lambda/dV_j$ tends to zero, The ratio $Cd\lambda/dV_j$, which is easier to handle numerically, can be defined as a voltage stability index for the entire system. In [20] the minimum real part of the eigenvalue of the Jacobian and in [6], the minimum singular value of the Jacobian are defined as voltage stability indexes. Fig.4.3 shows the variation of voltage stability index from the tangent vector.   The next section explains how to calculate the sensitivities of key components from the tangent vector.

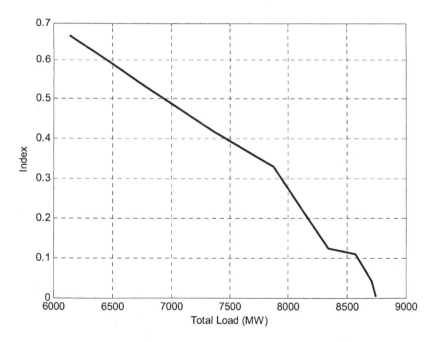

Fig.4.3 Variation voltage stability index with load

4.4.7 Sensitivity analysis from the tangent vector

We can derive an expression for the differential change in a scalar valued function $h(x, \lambda)$, with respect to differential change in λ. Here, $h(x, \lambda)$ is any power system operating constraint such as branch flow, reactive output of a generator, or bus voltage magnitude. Using the differential chain rule, we obtain

$$\frac{dh}{d\lambda} = \frac{\partial h}{\partial x}\frac{dx}{d\lambda} + \frac{\partial h}{\partial \lambda} \tag{4.3}$$

If we look closely at the Eq.4.3, for calculating $\frac{dh}{d\lambda}$ we need $\frac{dx}{d\lambda}$ (here x is the vector of voltage angles and magnitudes). This calculation involves computing the inverse of the Jacobian ($\frac{dx}{d\lambda} = -[F_x]^{-1}F_\lambda$) and fails at the critical point, where the Jacobian is singular and the inverse does not exist. But this $\frac{dx}{d\lambda}$ is given directly by the tangent vector in the continuation power flow. It can be substituted directly in Eq.4.3 to obtain the sensitivity of any operating constraint. The next section derives operating constraint sensitivities corresponding to load buses, branches, and generators [21].

4.4.8 Bus sensitivities

For bus sensitivities, the function $h(x, \lambda)$ can be either bus voltage magnitude or angle at a particular bus i. From Eq.4.3

$$\frac{dh}{d\lambda} = \sum_{j=1}^{n} \frac{\partial V_i}{\partial x_j}\frac{dx_j}{d\lambda} + \frac{\partial V_i}{\partial \lambda} = \frac{\partial V_i}{\partial V_i}\frac{dV_i}{d\lambda} + 0 = \frac{dV_i}{d\lambda} \tag{4.4}$$

Similarly, if we take bus voltage angle as function h, then

$$\frac{dh}{d\lambda} = \frac{d\delta_i}{d\lambda} \tag{4.5}$$

Close observation of the right-hand sides of the above two equations indicates that the numerators are nothing but the tangent vector elements. Because the value of $d\lambda$ is the same for each dV_i or $d\delta_i$ in a given tangent vector, bus sensitivities are nothing but the tangent vector elements themselves. Bus sensitivities indicate how weak a particular bus is near the critical point and help determine the areas close to voltage instability. The greater the bus sensitivity value, the weaker the bus is. Table 4.1 shows the bus sensitivities near the critical point, both according voltage angle and magnitude for the 39-bus test system.

Table 4.1 Bus sensitivities for the first 10 buses near the critical point – 39-bus system

According to voltage angle		
Bus number	Tangent Vector Element	Sensitivity
15	-0.17056	1.0000
20	-0.17026	0.99824
16	-0.14893	0.87318
18	-0.14372	0.84263
24	-0.14315	0.83929
8	-0.14226	0.83407
17	-0.1394	0.81730
9	-0.13504	0.79174
7	-0.13373	0.78406
39	-0.13275	0.77831

According to voltage magnitude		
Bus number	Tangent Vector Element	Sensitivity
23	-1.0000	1.0000
22	-0.98764	0.98764
21	-0.96496	0.96496
19	-0.92202	0.92202
20	-0.90683	0.90683
24	-0.89993	0.89993
16	-0.86676	0.86676
15	-0.83222	0.83222
17	-0.71916	0.71916
18	-0.65405	0.65405

4.4.9 Branch sensitivities

Let us consider a branch, ij. Let $V_i \angle \delta_i$ and $V_j \angle \delta_j$ be the voltage at buses i and j, respectively, and let $y_{ij} \angle \theta_{ij}$ be the line admittance. Then the losses in the line ij, neglecting the shunt charging capacitance, can be derived as follows.

The current in the branch ij is

$$I_{ij} = (V_i \angle \delta_i - V_j \angle \delta_j) y_{ij} \angle \theta_{ij}$$

(4.6)

Therefore,

$$I_{ij}^* = [(V_i \cos(\delta_i) - V_j \cos(\delta_j)) - j(V_i \sin(\delta_i) - V_j \sin(\delta_j))]$$

(4.7)

Then the total power flow in the branch ij from i to j is given by

$$(S_{loss})_{ij} = (P_{loss})_{ij} + j(Q_{loss})_{ij} = V_i I_{ij}^*$$
$$= [V_i^2 - V_i V_j \cos(\delta_i - \delta_j) - j V_i V_j \sin(\delta_i - \delta_j)] y_{ij} \angle - \theta_{ij}$$

(4.8)

Similarly, the total power flow from j to i is

$$(S_{loss})_{ji} = (P_{loss})_{ji} + j(Q_{loss})_{ji} = V_j I_{ji}^*$$
$$= [V_j^2 - V_i V_j \cos(\delta_j - \delta_i) - j V_i V_j \sin(\delta_j - \delta_i)] y_{ij} \angle - \theta_{ij}$$

(4.9)

The power loss in the branch ij is the algebraic sum of the above two power flows which is

$$P_{loss} + j Q_{loss} = [V_i^2 + V_j^2 - 2 V_i V_j \cos(\delta_j - \delta_i)] y_{ij} \angle - \theta_{ij}$$

(4.10)

Now differentiating the above loss function w.r.t. λ, we obtain the sensitivity equation as

$$\frac{dh}{d\lambda} = [(2V_i - 2V_j \cos(\delta_i - \delta_j)) \frac{dV_i}{d\lambda} + (2V_j - 2V_i \cos(\delta_i - \delta_j)) \frac{dV_j}{d\lambda}$$

(4.11)

$$+ (2 V_i V_j \sin(\delta_i - \delta_j)) \frac{d\delta_i}{d\lambda} - (2 V_i V_j \sin(\delta_i - \delta_j)) \frac{d\delta_j}{d\lambda}] y_{ij} \angle - \theta_{ij}$$

$$P_{loss} + j Q_{loss} = [V_i^2 + V_j^2 - 2 V_i V_j \cos(\delta_i - \delta_j)] y_{ij} \angle - \theta_{ij}$$

(4.12)

Defining this loss expression as function h, we obtain the sensitivity equation:

$$\frac{dh}{d\lambda} = [(2V_i - 2V_j \cos(\delta_i - \delta_j)) \frac{dV_i}{d\lambda} + (2V_j - 2V_i \cos(\delta_i - \delta_j)) \cdot \quad (4.13)$$

$$+ (2V_iV_j \sin(\delta_i - \delta_j)) \frac{d\delta_i}{d\lambda} - (2V_iV_j \sin(\delta_i - \delta_j)) \frac{d\delta_j}{d\lambda}]$$

$$y_{ij} \angle -\theta_{ij}$$

Branch sensitivity indicates how important a particular branch is to voltage stability. Table 4.2 shows the branch sensitivities obtained near the critical point by considering the Q_{losses} in the branches for the 39-bus test system. Fig.4.4 shows total Q_{losses} versus the real power for ten participating branches (five most and five least). These are the five branches with the highest and the lowest sensitivities. The slope of the curve showing the Q_{losses} in the five most participating branches is steep, compared with that of the five least participating branches. Thus, the rate at which the Q_{loss} in a particular branch is changing is important. This relation can be observed from the Q_{losses} and the sensitivities from Table 4.2. For example, in Table 4.2, the Q_{loss} in branch 37 is greater than that in branch 46, but branch 46 has a higher sensitivity.

Table 4.2 Branch sensitivities near the critical point for the 39-bus system

Branch no.	Bus i-Bus j	Q losses	Sensitivity
36	6-31	5.0039	1.0000
46	29-38	2.3526	0.7013
44	23-36	2.2039	0.4128
37	10-32	2.4342	0.3643
33	26-29	0.9644	0.3340
45	25-37	1.4208	0.3012
43	22-35	1.5627	0.2655
3	2-3	0.9236	0.2516
31	26-27	0.6612	0.2359
40	19-20	0.7395	0.2343

These sensitivities can provide information for contingency selection. In the 39-bus test system, branch 36, which has the highest sensitivity, is the most critical branch. This can be verified by considering the outage of each branch separately. With the outage of branch 36, we could transfer less power than we could with the outage of either branch 46 or 44, a fact indicating that branch 36 is more critical than branch 46 for voltage stability.

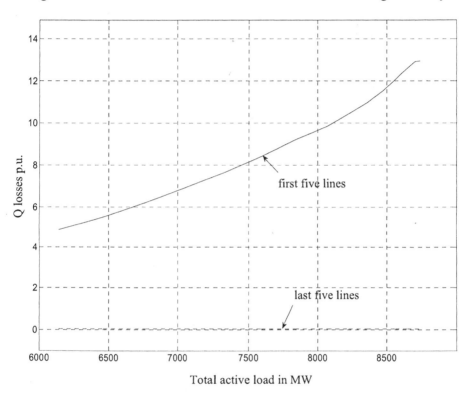

Fig.4.4 Q_{loss} vs. Real power

4.4.10 Generator sensitivities

The reactive power at a generator can be defined as the function h. i.e.,

$$h(x, \lambda) = Q_{Li0} + \lambda[K_{Li}S_{\Delta BASE}\sin(\psi_i)] + Q_{Ti} \qquad (4.14)$$

where

$$Q_{Ti} = \sum_{j=1}^{n} V_i V_j \sin(\delta_i - \delta_j - \theta_{ij})$$

with the following definitions:

- Q_{Li0} : original reactive load
- K_{Li} : multiplier designating the rate of load change at bus i as λ changes
- ψ_i : power factor angle of load change at bus i
- $S_{\Delta BASE}$: apparent power chosen to provide appropriate scaling of λ

The sensitivity equation therefore becomes

$$\frac{dh}{d\lambda} = \sum_{j=1}^{n} \frac{\partial Q_{Ti}}{\partial x_j} \frac{dx_j}{d\lambda} + K_{Li} S_{\Delta BASE} \sin(\psi_i) \tag{4.15}$$

Table 4.3 shows the generator sensitivities calculated for the same 39-bus test system near the critical point. There are 9 generators in the 30-bus system. Near the critical case only two generators are participating, and the other seven generators already have reached their $Q_{\lim its}$. Generator sensitivities indicate those generators that are important in maintaining voltage stability near the critical point. Evidently, generators with high sensitivity are especially important. These generator sensitivities can be used to obtain a better combination of generators to share the increase in load. Sensitivity results can be verified with the finite difference approach.

The sensitivities discussed in the above sections are useful not only for finding weak areas in the system, but also for diagnosing modeling deficiencies. In [22], an analysis of voltage stability on the MAPP-MAIN transmission interface of Wisconsin used three different power flow models to determine voltage stability limits. They used bus and branch sensitivities of CPF to assess and compare modeling deficiencies or strengths for three types of power flow models they used. They also used branch sensitivities to identify the most critical branches in the system, both for normal case and for some contingencies.

Table 4.3 Generator sensitivities near the critical point for the 39-bus system

Generator No.	Q generation	Sensitivity
39	4.6949	1.0000
37	2.7001	0.9278
38	4.6477	0.7666

4.4.11 Qualitative vs. quantitative sensitivities

Qualitative sensitivity refers to the fact that it only gives direction and relative magnitude of change of stability indices under parameter variations. Quantitative sensitivity can be used for the quantification of change of the stability index with respect to a change of some parameter. A good example of qualitative sensitivity is eigenvalue sensitivity. At an operating point, $\partial \lambda_i / \partial p$ gives qualitative information about the parameter's influence on the eigenvalue. Because the eigenvalue is a highly nonlinear function of system parameters, it is practically impossible to quantitatively estimate the change in the eigenvalue due to variations of some parameters.

Similarly the sensitivities based on tangent vetor elements and modal analyses are also providing qualitative information.

As mentioned in section 4.2, large deviation based indices account for the nonlinearities cased by larger disturbances. Since these indices are usually defined in the load power space, they characterize the critical operating condition from a parameter space point of view.   To get the qualitative as well as quantitative margin information, one should derive the sensitivity of the critical point for any given change of parameters. Next sections provide the details related to margin sensitivity.

4.5 Margin Sensitivity

As introduced in chapter 3, a scalar λ denotes the system load/generation level is called the bifurcation parameter. The system reaches a state of voltage collapse, when λ hits its maximum value (the turning point). For this reason, the system DAE model at equilibrium state is parameterized by this bifurcation parameter λ. When system parameters are changed, the total transfer capability will probably increase or decrease. References [23, 24] first derived this change of margin for changing parameters. The change of transfer margin can be determined if the change of λ between two bifurcation points on the voltage collapse boundary Σ is known. Since we are interested in estimating the loading margin when some arbitrary parameters are varied, we rewrite the DAE as follows to denote the parameter dependence of the system solution.

$$\dot{X} = F(X(\lambda(P), P), Y(\lambda(P), P), \lambda(P), P) \qquad (4.16)$$

$$0 = G(X(\lambda(P), P), Y(\lambda(P), P), \lambda(P), P) \qquad (4.17)$$

At a saddle node bifurcation, which is also an equilibrium point (though not asymptotically stable), we take the partial differentiation of the above two equations with respect to the parameter vector P; then

$$0 = \frac{\partial F}{\partial X}\left(\frac{\partial X}{\partial \lambda}\frac{\partial \lambda}{\partial P} + \frac{\partial X}{\partial P}\right) + \frac{\partial F}{\partial Y}\left(\frac{\partial Y}{\partial \lambda}\frac{\partial \lambda}{\partial P} + \frac{\partial Y}{\partial P}\right) + \frac{\partial F}{\partial \lambda}\frac{\partial \lambda}{\partial P} + \frac{\partial F}{\partial P} \quad (4.18)$$

$$0 = \frac{\partial G}{\partial X}\left(\frac{\partial X}{\partial \lambda}\frac{\partial \lambda}{\partial P} + \frac{\partial X}{\partial P}\right) + \frac{\partial G}{\partial Y}\left(\frac{\partial Y}{\partial \lambda}\frac{\partial \lambda}{\partial P} + \frac{\partial Y}{\partial P}\right) + \frac{\partial G}{\partial \lambda}\frac{\partial \lambda}{\partial P} + \frac{\partial G}{\partial P} \quad (4.19)$$

Simplifying the above expressions gives

$$\begin{pmatrix} F_X & F_Y \\ G_X & G_Y \end{pmatrix}\begin{pmatrix} \frac{\partial X}{\partial \lambda}\frac{\partial \lambda}{\partial P} + \frac{\partial X}{\partial P} \\ \frac{\partial Y}{\partial \lambda}\frac{\partial \lambda}{\partial P} + \frac{\partial Y}{\partial P} \end{pmatrix} + \begin{pmatrix} F_\lambda \\ G_\lambda \end{pmatrix}\frac{\partial \alpha}{\partial P} + \begin{pmatrix} F_P \\ G_P \end{pmatrix} = 0 \qquad (4.20)$$

Premultiplying the above equation by (v_F^T, v_G^T) corresponding to the zero eigenvalue of A_{sys}.

For the zero eigenvalues of A_{sys}

$$(v_F^T, v_G^T)\begin{pmatrix} F_X & F_Y \\ G_X & G_Y \end{pmatrix} = \begin{bmatrix} 0 & 0 \end{bmatrix}$$

So the first term in Eq.4.20 will vanish. Then

$$(v_F^T, v_G^T)\begin{pmatrix} F_\lambda \\ G_\lambda \end{pmatrix}\frac{\partial \lambda}{\partial P} + (v_F^T, v_G^T)\begin{pmatrix} F_P \\ G_P \end{pmatrix} = 0 \qquad (4.21)$$

Therefore, the bifurcation parameter sensitivity is [23, 24]:

$$\frac{\partial \lambda}{\partial P} = \frac{(v_F^T, v_G^T)\begin{pmatrix} F_P \\ G_P \end{pmatrix}}{(v_F^T, v_G^T)\begin{pmatrix} F_\lambda \\ G_\lambda \end{pmatrix}} \qquad (4.22)$$

Using vector notation (the underline sign), we can write the generation and load parameterization equations as

$$\underline{P}_{Gs} = \left(\cdots P_{Gsi}(\lambda) \cdots \right)^T \tag{4.23}$$

$$\underline{L}(\lambda) = \left(\cdots P_{Li}(\lambda) \cdots Q_{Li}(\lambda) \right)^T \tag{4.24}$$
$$= \underline{L}_0 + \lambda \underline{K}$$
$$= \left(\underline{P}_L \quad \underline{Q}_L \right)^T$$

where

$$\underline{K} = \left(\cdots K_{Lp_i} P_{Li_0} \cdots K_{Lq_i} Q_{Li_0} \cdots \right)^T \tag{4.25}$$
$$= \left(\cdots K_{P_i} \cdots K_{Q_i} \cdots \right)^T$$
$$= \left(\underline{K}_P \quad \underline{K}_Q \right)^T \tag{4.26}$$

denotes the loading pattern. At saddle node bifurcation, the set of real powers form the voltage instability boundary in the load power space. Using these notations and noting that only L contains λ, by the chain rule, the derivatives of F and G with respect to λ can be written as

$$\begin{pmatrix} \frac{\partial F}{\partial \lambda} \\ \frac{\partial G}{\partial \lambda} \end{pmatrix} = \begin{pmatrix} \frac{\partial F}{\partial L} \\ \frac{\partial G}{\partial L} \end{pmatrix} \underline{K} \tag{4.27}$$

Substitution of (4.27) in (4.22) results in:

$$\frac{\partial \lambda}{\partial P} = -\frac{(v_F^T, v_G^T) \begin{pmatrix} F_P \\ G_P \end{pmatrix}}{(v_F^T, v_G^T) \begin{pmatrix} F_\lambda \\ G_\lambda \end{pmatrix}}$$

$$= -\frac{(v_F^T, v_G^T) \begin{pmatrix} F_P \\ G_P \end{pmatrix}}{(v_F^T, v_G^T) \begin{pmatrix} F_L \\ G_L \end{pmatrix} \underline{K}} \tag{4.28}$$

This margin sensitivity gives the first order partial derivative in the Taylor series expansion of λ as a nonlinear function of P, which describes the hypersurface Σ.

The bifurcation parameter sensitivity will allow us to know, when some parameters are varied, how the system will move along the hypersurface Σ in the vicinity of the current instability point denoted by λ_*.

Reference [25] explained margin sensitivity [23] in the framework of DAE formulation. Invariance subspace parametric sensitivity in the context of voltage stability is discussed in [26].

4.5.1 Transfer margin estimation

Once $\partial\lambda/\partial P$ is computed, we will first get the bifurcation parameter estimation as

$$\Delta\lambda = \frac{\partial\lambda}{\partial P}\Delta P \tag{4.29}$$

where P contains all the parameters explicitly appearing in the DAE model including the load scenario parameters. If we are only interested in the real power transfer capability, then we define $PLM(\lambda_*)$ as the total power of all the buses at voltage collapse before a parameter variation, and $PLM(\lambda_*')$ as the total power of all the buses at voltage collapse after a parameter variation. In the case of a non-real-power-load related parameter, we will get the margin change estimate as

$$\Delta PLM = PLM(\lambda_*') - PLM(\lambda_*) \tag{4.30}$$

$$\triangleq \sum_{i=1}^{N} \Delta P_{L_i}(\lambda)$$

$$= \Delta\lambda \sum_{i=1}^{N} K_{P_i}$$

And, the new critical powers at all the buses, in vector notation, can be estimated as

$$\underline{P_L}'_* = \underline{P_L}_* + \Delta\lambda\underline{K}_P \tag{4.31}$$

The above discussion is conceptually illustrated in Fig.4.5 and Fig.4.6.
In the case of a real power load related parameter (K_{Lpi} and P_{Li0}) variation, the loading margin estimation for bus will be:

$$\Delta P_{L_i}(\lambda, K_{Lpi}, P_{Li0}) = \Delta\lambda\underline{K}_{Pi} + \frac{\partial P_{L_i}}{\partial K_{Lpi}}\Big|_* \Delta K_{Lpi} + \frac{\partial P_{L_i}}{\partial P_{Li0}}\Big|_* \Delta P_{Li0} \quad (4.32)$$

where

$$\frac{\partial P_{L_i}}{\partial K_{Lpi}}\Big|_* = \lambda_* P_{Li0} \quad (4.33)$$

$$\frac{\partial P_{L_i}}{\partial P_{Lp0}}\Big|_* = 1 + \lambda_* K_{Lpi} \quad (4.34)$$

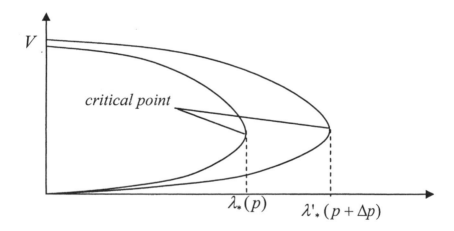

Fig. 4.5 Transfer margin as shown on a PV curve

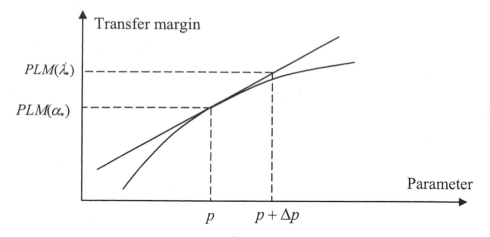

Fig. 4.6 Transfer margin estimation

The total margin change estimation will then be modified to include two more terms

$$\Delta PLM = \sum_{i=1}^{N} \Delta P_{L_i} = \Delta\lambda \sum_{i=1}^{N} \underline{K}_{Pi} + \sum_{i \in I_{ch}} \frac{\partial P_{L_i}}{\partial K_{Lpi}}|_* \Delta K_{Lpi} + \sum_{i \in I_{ch}} \frac{\partial P_{L_i}}{\partial P_{Li0}} \quad (4.35)$$

where I_{ch} denotes the set which includes all the buses under load parameter variations. When reactive power load parameters (K_{Lqi} and Q_{Li0}) are varied, the real power transfer margin estimation can still be calculated by using Eqs.4.30 and 4.31. The reactive power transfer margin estimation, however, should be made by using equations similar to Eqs.4.32 and 4.35 for bus i and the total.

4.5.2 Multi-parameter margin sensitivity

In modern power system operation, coordinated controls are often used to optimize certain performance indices, for instance, to maximize the transfer on a specified transmission interface if possible. Since a first order estimation can be linearly superimposed, we can study the combined parameter influence on stability margin variation by using

$$\Delta \lambda = \frac{\partial \lambda}{\partial P_1} \Delta P_1 + \cdots + \frac{\partial \lambda}{\partial P_m} \Delta P_m \tag{4.36}$$

However, when more than one parameter is varied, the mixed partial derivative term of higher orders also contributes to the margin variation. For instance, when both p_i and p_j are varied, the mixed second order term $\frac{\partial^2 \alpha}{\partial p_i \partial p_j} \Delta p_i \Delta p_j$ is also nonzero. Inaccuracy will result from ignoring this term in addition to $\frac{\partial^2 \alpha}{\partial p_i^2} \Delta p_i^2$.

4.5.3 Sensitivity formulas

This subsection will provide the sensitivity formulas with respect to the parameters studied.

- Sensitivity matrices $\frac{\partial F}{\partial P}$ and $\frac{\partial G}{\partial P}$

Excitation system parameters

 - Exciter gain K_{ai}:

$$\frac{\partial f_{g8i}}{\partial K_{ai}} = \frac{1}{T_{ai}} (V_{refii} - V_i - R_{fi}) \tag{4.37}$$

where f_{g8i} is the right hand side of Eq.3.32

 - Self excitation parameter K_{ei}:

$$\frac{\partial f_{g7i}}{\partial K_{ei}} = -\frac{E_{fdi}}{T_{ei}} \tag{4.38}$$

$$\frac{\partial f_{g9i}}{\partial K_{ei}} = -\frac{K_{fi} E_{fdi}}{T_{ei} T_{fi}} \tag{4.39}$$

where f_{g7i} and f_{g9i} are the right hand sides of Eqs.3.31 and 3.33 respectively

– Exciter reference voltage V_{refi}:

$$\frac{\partial f_{g8i}}{\partial V_{refi}} = \frac{K_{ai}}{T_{ai}} \tag{4.40}$$

Governor parameters

– Governor base case setting P_{Gsi0}:

$$\frac{\partial f_{g6i}}{\partial P_{gsi0}} = \frac{1}{T_{gi}} \tag{4.41}$$

where f_{g6i} is the right hand side of Eq.3.35.

Network parameters

– Line susceptance B_{ij}:

$$\frac{\partial \Delta P_i}{\partial B_{ij}} = -\frac{\partial P_{Ti}}{\partial B_{ij}}$$

$$= -\frac{\partial}{\partial B_{ij}} (V_i \sum_{k=1}^{N} V_k y_{ik} \cos(\theta_i - \theta_k - \gamma_{ik}))$$

$$= V_i V_j \sin(-\theta_i + \theta_j) \tag{4.42}$$

$$\frac{\partial \Delta Q_i}{\partial B_{ij}} = -\frac{\partial Q_{Ti}}{\partial B_{ij}}$$

$$= -\frac{\partial}{\partial B_{ij}} (V_i \sum_{k=1}^{N} V_k y_{ik} \sin(\theta_i - \theta_k - \gamma_{ik}))$$

$$= V_i V_j \cos(\theta_i - \theta_j) \tag{4.43}$$

where ΔP_i and ΔQ_i are from the real and reactive power mismatch equations as given in Eqs.3.16 and 3.17 .

– Shunt capacitance B_{i0}:

$$\frac{\partial \Delta Q_i}{\partial B_{i0}} = -\frac{\partial Q_{Ti}}{\partial B_{i0}}$$

$$= -\frac{\partial}{\partial B_{i0}}(V_i \sum_{k=1}^{N} V_k y_{ik} \sin(\theta_i - \theta_k - \gamma_{ik}))$$

$$= V_i^2 \qquad (4.44)$$

Load (scenario) parameters

– Real power load increase speed parameter K_{Lpi}:

$$\frac{\partial f_{g6i}}{\partial K_{Lpj}} = \frac{1}{T_{gi}} K_{Gpi} P_{Lj0} \qquad (4.45)$$

$$\frac{\partial \Delta P_i}{\partial K_{Lpi}} = -\lambda P_{Li0} \qquad (4.46)$$

– Reactive power load increase speed parameter K_{Lqi}:

$$\frac{\partial \Delta Q_i}{\partial K_{Lqi}} = -\lambda Q_{Li0} \qquad (4.47)$$

– Base case real power load P_{Li0}:

$$\frac{\partial f_{g6i}}{\partial P_{Lj0}} = \frac{1}{T_{gi}} \lambda K_{Gpi} K_{Lpj} \qquad (4.48)$$

$$\frac{\partial \Delta P_i}{\partial P_{Li0}} = -(1 + \lambda K_{Lpi}) \qquad (4.49)$$

– Base case reactive power load Q_{Li0}:

$$\frac{\partial \Delta Q_i}{\partial Q_{Li0}} = -(1 + \lambda K_{Lqi}) \qquad (4.50)$$

The above formulas are used to construct the sensitivity matrices $\partial F/\partial P$ and $\partial G/\partial P$.

– Sensitivity matrices $\partial F/\partial L$ and $\partial G/\partial L$:

$$\frac{\partial f_{g6i}}{\partial P_{Lj}(\lambda)} = \frac{K_{Gpi}}{T_{gi}} \qquad (4.51)$$

$$\frac{\partial \Delta P_i}{\partial P_{Li}(\lambda)} = -1.0 \qquad (4.52)$$

$$\frac{\partial \Delta Q_i}{\partial Q_{Li}(\lambda)} = -1.0 \qquad (4.53)$$

4.6 Test System Studies

In this section, the proposed sensitivity measure calculated at saddle node bifurcation is applied to estimate the voltage stability margin under system parameter variations. Physical interpretations are given following the test results. The method is demonstrated through two test system examples. First we provide a simple two bus example considered in previous chapters to demonstrate the steps involved in calculating these sensitivities.

4.6.1 Two bus example:

Consider the two bus example (Fig. 4.7) introduced previously:

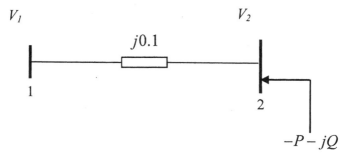

Fig. 4.7 Two bus system

With two axis model, there are totally 9 states. The differential and algebraic equations describing the system are given below.

F part:

$$\dot{\delta}_1 = (\omega_1 - \omega_m)\omega_0$$

$$\dot{\omega}_1 = M_1^{-1}[P_{m1} - D_1(\omega_1 - \omega_m) - (E'_{q1} - X'_{d1}I_{d1})I_{q1}$$
$$- (E'_{d1} + X'_{q1}I_{q1})I_{d1}]$$

$$\dot{E}'_{q1} = T_{d01}^{-1}[E_{fd1} - E'_{q1} - (X_{d1} - X'_{d1})I_{d1}]$$

$$\dot{E}'_{d1} = T_{q01}^{-1}[-E'_{d1} + (X_{q1} - X'_{q1})I_{q1}]$$

$$I_{d1} = [R_{s1}E'_{d1} + E'_{q1}X'_{q1} - R_{s1}V_1\sin(\delta_1 - \theta_1) - X'_{q1}V_1\cos(\delta_1 - \theta_1)]A_1^{-1}$$

$$I_{q1} = [R_{s1}E'_{q1} - E'_{d1}X'_{d1} - R_{s1}V_1\cos(\delta_1 - \theta_1) - X'_{d1}V_1\sin(\delta_1 - \theta_1)]A_1^{-1}$$

$$A_1 = R_{s1}^2 + X'_{d1}X'_{q1}$$

$$\dot{E}_{fd1} = T_{e1}^{-1}[V_{r1} - [S_{e1}(E_{fd1})]E_{fd1}]$$

$$\dot{V}_{r1} = T_{a1}^{-1}[-V_{r1} + K_{a1}(V_{ref1} - V_1 - R_{f1})]$$

If $V_{r1,min} \le V_{r1} \le V_{r1,max}$, $V_{pss1} = 0$ (at steady state),

$$\dot{R}_{f1} = T_{f1}^{-1}[-R_{f1} - [K_{e1} + S_{e1}(E_{fd1})]K_{f1}E_{fd1}/T_{e1} + K_{f1}V_{r1}/T_{e1}]$$

$$\dot{P}_{m1} = T_{ch1}^{-1}(\mu_1 - P_{m1})$$

$$\dot{\mu}_1 = T_{g1}^{-1}[P_{gs1} - (\omega_1 - \omega_{ref})/R_1 - \mu_1] \qquad \text{if } \mu_{1,min} \le \mu_1 \le \mu_{1,max}$$

$$P_{gs1} = P_{gs1}^0 (1 + K_{g1}\lambda)$$

G part:

$$0 = P_{g1} - P_{t1}$$
$$0 = Q_{g1} - Q_{t1}$$
$$0 = 0 - P_{l20}(1 + \lambda K_{lp2}) - P_{t2}$$
$$0 = 0 - Q_{l20}(1 + \lambda K_{lq2}) - Q_{t2}$$

where,

$$\begin{cases} P_{g1} = I_{d1}V_1 \sin(\delta_1 - \theta_1) + I_{q1}V_1 \cos(\delta_1 - \theta_1) \\ Q_{g1} = I_{d1}V_1 \cos(\delta_1 - \theta_1) - I_{q1}V_1 \sin(\delta_1 - \theta_1) \end{cases}$$

$$\begin{cases} P_{t2} = Y_{21}V_1V_2 \cos(\theta_2 - \theta_1 - \varphi_{21}) + Y_{22}V_2^2 \cos(\varphi_{22}) \\ Q_{t2} = -Y_{21}V_1V_2 \sin(\theta_2 - \theta_1 - \varphi_{21}) - Y_{22}V_2^2 \sin(\varphi_{22}) \end{cases}$$

$$|Y_{21}| = 10, \varphi_{21} = 90°, \varphi_{22} = -90°$$

Parameters used in this example are given in Appendix A.
Given the initial operating condition,

$$P_{l20} = 1.4\,p.u., Q_{l20} = 0, P_{Gs1}^0 = 1.4\,p.u., \lambda = 0, K_{g1} = 1.0, K_{lp2} = 1.0$$

The Newton method is first used to reach a viable operating point. Then
the EQTP program is used to get the critical point, where the left hand side
of the F part equals to zero.

From Eq.4.28,

$$\frac{\partial \lambda}{\partial P} = -\frac{(v_F^T, v_G^T)\begin{pmatrix} F_P \\ G_P \end{pmatrix}}{(v_F^T, v_G^T)\begin{pmatrix} F_\lambda \\ G_\lambda \end{pmatrix}}$$

Consider the margin sensitivity with respect to shunt capacitance B_{20} at bus 2. Because B_{20} is contained only in the power flow equations (Y_{22}),

$F_P = \frac{\partial F}{\partial B_{20}} = 0$, $G_P = \frac{\partial G}{\partial B_{20}} = V_2^2$ (for G_P vector all the elements are zero except one.   For the F_P vector all the elements are zero). Hence

$$(v_F^T, v_G^T)\begin{pmatrix} F_P \\ G_P \end{pmatrix} = v_G^T G_P = -0.13 * 0.6634 * 0.6634 = -0.0572$$

$$\underline{P_{Gs}} = \begin{pmatrix} P_{Gs1}^0(1 + K_{g1}\lambda) & 0 \end{pmatrix}^T$$

$$\underline{L(\lambda)} = \begin{pmatrix} 0 & P_{l20}(1 + \lambda K_{lp2}) & 0 & 0 \end{pmatrix}^T$$
$$= \underline{L_0} + \lambda\underline{K}$$
$$= \begin{pmatrix} \underline{P_L} & \underline{Q_L} \end{pmatrix}^T$$

$$\underline{K} = \begin{pmatrix} 0 & K_{lp2}P_{l20} & 0 & 0 \end{pmatrix}^T$$
$$= \begin{pmatrix} \underline{K_P} & \underline{K_Q} \end{pmatrix}^T$$

Similarly

$$(v_F^T, v_G^T)\begin{pmatrix} F_\lambda \\ G_\lambda \end{pmatrix} = v_F^T F_\lambda + v_G^T G_\lambda = v_F^T T_{g1}^{-1} P_{gs1}^0 K_{g1} + v_G^T K_{lp2}P_{l20} = 0.2447$$

Then

$$\frac{\partial\lambda}{\partial P} = -\frac{(v_F^T, v_G^T)\begin{pmatrix} F_P \\ G_P \end{pmatrix}}{(v_F^T, v_G^T)\begin{pmatrix} F_\lambda \\ G_\lambda \end{pmatrix}} = -\frac{-0.0572}{0.2447} = 0.2338$$

Suppose there is a 0.1 p.u. shunt capacitance change at bus 2, then from Eq.4.29

$$\Delta\lambda = \frac{\partial\lambda}{\partial B_{20}}\Delta B_{20} = 0.2338*0.1 = 0.2338 \ p.u.$$

$$\Delta PLM = \Delta\lambda\sum_{i=1}^{N} K_{P_i} = 0.2338*1.4 = 0.03273 p.u. = 3.273 MW$$

At base case, the load is 140 MW. The critical point (with unit power factor load increase) occurs at 310.2743 MW. Hence, the estimated new loadability will be: 310.2743 + 3.273 = 313.5473 MW. The actual loadability with EQTP tracing is 313.7188 MW.

The margin sensitivity method works quite well for this case.

The P-V curve (the abscissa is the normalized value of real power, where $p = PX/E^2$) before and after the adding of shunt capacitance is shown in Fig.4.8:

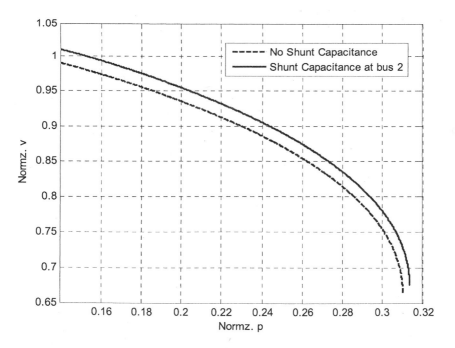

Fig.4.8   PV curve with and without shunt capacitance

If we consider only power flow formulation then this case corresponds to unity power factor and the corresponding margin will be 500MW for base case and 505 MW with an increase of 0.1 p.u shunt capacitance at bus2.

Fig. 4.9 shows the linear rage of the margin sensitivity.

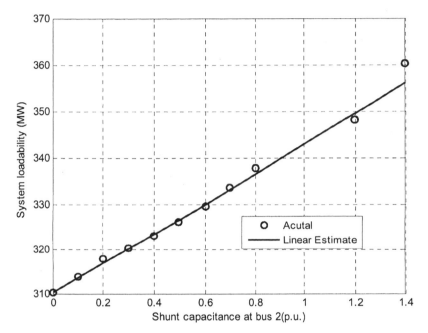

Fig.4.9 System loadability vs. shunt capacitance

4.6.2 The New England system

Scenario1 setting in chapter 3 was used to locate the saddle node bifurcation point. With the nominal parameter settings, the total real power transfer margin between the base case and the critical point is approximately 2738 MW.

The following sections provide the margin sensitivity of various parameters.

4.6.2.1 Exciter parameters

The automatic voltage regulator (Fig.4.10) of the generator at bus 39 affects the voltage collapse limited transfer the most. Obviously, in calibration (parameter estimation), K_{a39} should be given first priority for better accuracy. Otherwise the voltage stability limited transfer evaluation might be in error.

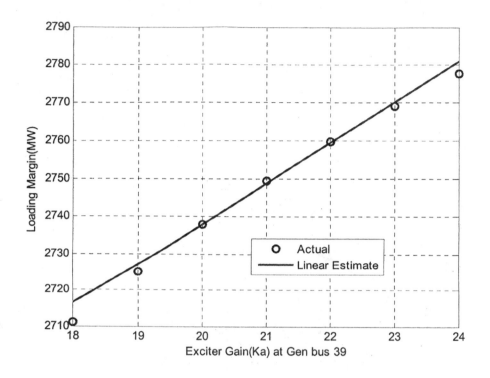

Fig. 4.10 Loading margin vs. exciter gain

The exciter reference voltage (Fig.4.11) is one of the control settings that can be used to control the generator terminal voltages. From Fig.4.11, we can see that V_{ref39} is very effective for the increase of the transfer.

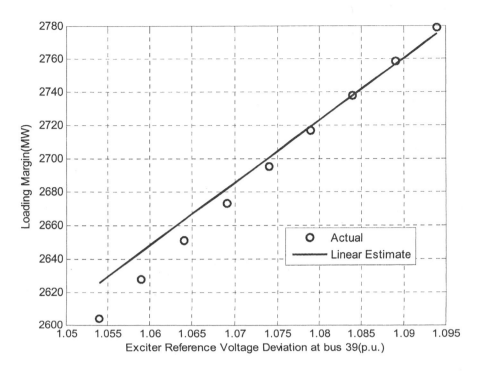

Fig. 4.11 Margin vs. exciter reference voltage

4.6.2.2 Network parameters

One of the reasons for voltage instability is the lack of reactive power support at critical locations [9]. Supplying enough VARs (Fig.4.12) locally at or near heavily loaded buses, or at an intermediate point between generation and load centers usually increases the real power transfer capability. In this test case, according to the sensitivity, bus 10 is one of the best places to put some reactive power support. Bus 10 is linked to the generator bus 32 where this generator is at its limit. It is shown in Fig.4.12 that a 0.25 p.u shunt capacitance installation will lead to an increase of approximately 44 MW in total real power transfer. The linear estimate is very accurate over a wide range of shunt values. Selecting the best location for installation of SVC can be analyzed using the same information.

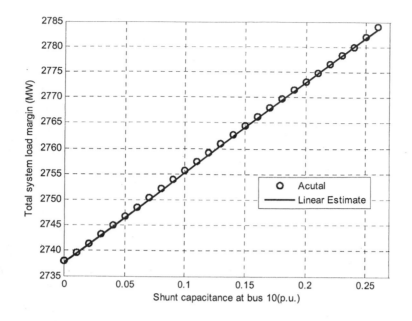

Fig. 4.12 Loading margin vs. shunt capacitance

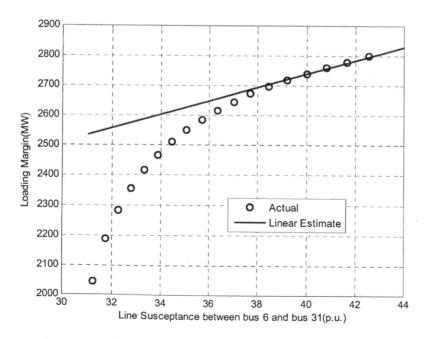

Fig. 4.13 Loading margin vs. line susceptance

Line susceptance (Fig.4.13) is also a critical parameter in transfer capability evaluation. One of the most sensitive lines indicated by margin sensitivity is line 6-31. Since the generator at bus 31 is one of the remaining generation sources not hitting the limits, transfer capability can be increased by reducing the reactance of that line. This will enable the network to receive more reactive power from the generator at bus 31. The margin curve becomes nonlinear when ΔB_{6-31} exceeds 3 p.u.(over 7% of its nominal value. Nominal value is 40 p.u.). Quantitative study of the influence of line susceptances on transmission capability can be extended to analyze the effectiveness of FACTs devices, such as that of TCSC (thyristor controlled series capacitors). Line contingency could also be simulated through this [24].

4.6.2.3 Load (scenario) parameters

The parameters K_{Lpi} and K_{Lqi} designate the rate of load increase at bus i. If they are zero, the loads will remain at the base case value. By changing these load scenario parameters, the power factors will be varied. At the nominal case, we give 1.0 to both K_{Lpi}'s and K_{Lqi}'s. This will force the load to increase at a constant power factor. For the current scenario, margin sensitivity (Fig.4.14) indicates that K_{Lp20} is the most sensitive. Forcing the load at bus 20 to remain unchanged (giving 0 to K_{Lp20}) will increase the loading margin of the remaining buses. However, since the load at bus 20 is very large, the overall margin will decrease. For the same reason, when we increase the load at this bus at a faster rate by giving K_{Lp20} a value larger than 1.0, the total margin does not increase significantly, rather saturation occurs. Therefore, the linear sensitivity does not work well in this case.

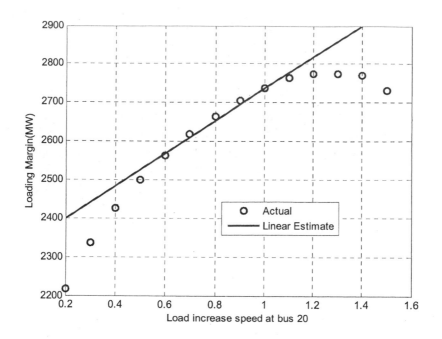

Fig. 4.14 Loading margin vs. load parameter K_{Lpi}

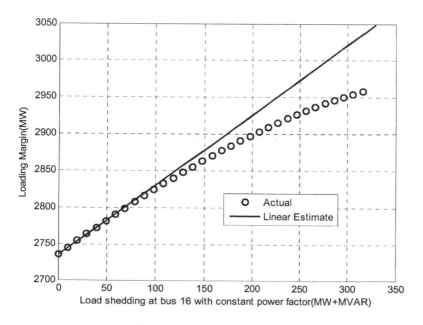

Fig. 4.15 Loading margin vs. load shedding

The base case real and reactive power loads at bus 16 are 329.4 MW and 132.3 MVAR respectively. If we shed up to 1.0 p.u. load at a constant power factor, the resultant transfer margin (the difference between the base case and the critical point) will increase almost linearly to about 2825 MW. The linear estimate (Fig.4.15) again works quite well near this region. But when more loads are shed, the margin curve becomes much nonlinear.

4.6.2.4 Multiple-parameter variations

Two parameters (shunt capacitance at bus 10 and exciter reference voltage at bus 39) are changed simultaneously. Eq.4.36 was used to predict the margin. In Fig. 4.16, the transfer margin is plotted against both of these two parameters. The linear prediction is very accurate over the range of parameter variations.

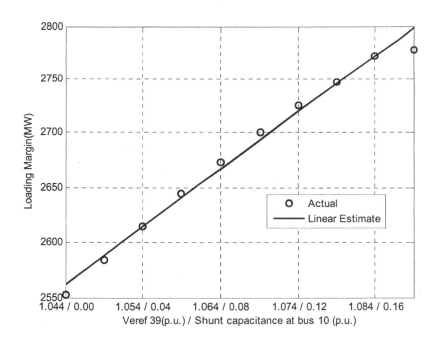

Fig. 4.16 Margin under multiarameter changes - B_{i010} & V_{REF39}^{EX}

References

[1]   Taylor, C., Indices predicating voltage collapse including dynamic phemomena. CIGRE Task Force 38.02.11, 1994

[2]   Cañizares, C.A.(Editor),Voltage Stability Assessment: Concepts, Practices and Tools: IEEE – PES Power Systems Stability Subcommittee Special Publication SP101PSS (ISBN 01780378695) , 2003

[3]   Venikov, S. A., Idelchick I., Tarasov, I., Estimation of electric power system steady-state stability in load flow calculation. IEEE Trans. Power Appar. Syst. 94, pp. 1034-1041, 1975

[4]   Hager, W. W., Applied numerical algebra, New Jersey: Prentice Hall Inc., 1988

[5]   Golub, G.H., Van, L.C.F., Matrix computations, Baltimore: The johns Hopkins university press, 1990

[6]   Lof, P.A., Anderson, G., Hill, D.J., Voltage stability indices for stressed power systems IEEE PES winter meeting New York, 1992

[7]   Tiranuchit, A., Thomas, R.J., A posturing strategy against volt age instabilities in electric power systems IEEE Trans. Power Systems 3: 87-93, 1988

[8]   Canizares, C. A., DeSouza, A. C. Z., Quintana, V. H., Comparison of performance indices for detection of proximity to voltage collapse IEEE PWR SM 583-5, 1995

[9]   Ajjarapu, V., Christy, C., The continuation power flow: A tool for steady state voltage stability analysis. IEEE Trans. Power Syst.7: 417-423, 1992

[10]   Chiang, H.D., Flueck, A.J., Shah, K.S., and Balu, N., "CPFLOW: a practical tool for tracing power system steady-state stationary behavior due to load and generation variations,"*IEEE Trans. on power systems*, PWRS-10, pp.623-634, May 1995

[11]   Flueck, A.J., Dondeti, J.R., "A new continuation power flow tool for investigating the nonlinear effects of transmission branch parameter variations,"*IEEE Trans. on power systems*, vol.15, no.1, pp.223-227, Feb. 2000

[12]   Iba, K., Suzuki, H., Egawa, M., Watanabe, T., *Calculation of critical loading condition with nose curve using homotopy continuation method*, IEEE Transactions on Power Systems, Volume 6, Issue 2, pp.584 – 593, May 1991

[13]   Van Cutsem, T., A method to compute the reactive power margins with respect to voltage collapse. IEEE   Trans. Power Syst.6:   145-156, 1991

[14] Ajjarapu, V., Identification of steady state voltage stability in power systems. Int. J. Energy Syst. 11: 43-46, 1991

[15] Canizares, C. A., Alvarado, F. L., Demarco, C. L., Dobson, I., Long, W. F., Point of collapse methods applied to ac/dc power systems. IEEE   Trans. Power Syst.7:   673-683, 1992

[16] Alvarado, F. L., Dobson, I., Hu, Y., Computation of closest bifurcations in power systems IEEE Trans. Power Syst.9: 918-928, 1994

[17] Overbye, T. J., Dobson, I., Demarco, C. L., Q-V curve interpretation of energy measure for voltage security. IEEE   Trans. Power Syst.9: 331-340, 1994

[18] Smed, T., Carr, J., Feasible eigenvalue sensitivity for large power systems. IEEE Trans. Power Syst.8: 555-563, 1981

[19] Gao, B., Morison, K., Kundur, P. , Voltage stability evaluation using modal analysis IEEE   Trans. Power Syst.7: 1529-1536, 1992

[20] Ilic, M., Stankovic, A., Voltage problems on transmission networks subject to unusual power flow patterns, IEEE Trans. Power Syst., vol.6, no.1, pp.339-348, Feb.1991

[21] Ajjarapu, V., Battula, S., Sensitivity based continuation power flow. Proceedings of the 24th Annual North American Power Symposium, University of Nevada, Reno, October, 1992

[22] Roddy, R. W., Christy, C. D., Voltage stability on the MAPP-MAIN transmission interface of Wisconsin. Proceedings of the EPRI NERC forum on voltage stability, Breckenridge , Colorado, 1992

[23] Greene, S., Dobson, I., Alvarado, F. L., Sensitivity of the loading margin to voltage collapse with respect to arbitrary parameters, IEEE Trans. Power Syst., vol.12, no.1,pp. 262-272, Feb.1997

[24] Greene, S., Dobson, I., Alvarado, F. L., Contingency ranking for voltage collapse via sensitivities from a single nose curve, IEEE Trans. Power Syst., vol.14, no.1, pp. 232-240, Feb.1999

[25] Long, B., Ajjarpau, V., The sparse formulation of ISPS and its application to voltage stability margin sensitivity and estimation. IEEE Trans. Power Syst., vol.14, pp. 944-951, August 1999

[26] Lee, B., Ajjarapu, V., Invariant subspace parametric sensitivity (ISPS) of structure preserving power system models IEEE Trans. Power Syst., vol.11, pp.845-850, May 1996

5 Voltage Stability Margin Boundary Tracing

In Chapter 4 we discussed about various sensitivities that can be used to identify factors that may lead to voltage instability. In general these sensitivities are valid within the narrow range of parameter variation. This chapter provides the methodologies that extend the range of these parameter variations. The system load margin corresponding to any control configuration can be determined without retracing the entire PV curve.

5.1 Introduction

Deregulation brings new challenges to operate the power system. Independent System Operator (ISO) needs to monitor the system load margin in a real time and close the power transaction deals based on the available system stability margin as well as other considerations to meet the quickly varying energy demand. How to efficiently extend the system margin by the readjustment of the system control configuration becomes an important part of the overall operation of the power system.

Contingency causes system margin to shrink and could endanger a system. Hence load margin variation with respect to specified contingency could be a security index that can be applied in contingency screening [1] and operation planning.

Margin boundary can be obtained in a variety of ways. The trivial way to obtain a new margin point is to retrace the PV curve with changed system conditions. Obviously this method is time consuming and less informative.

As discussed in chapter 4 the boundary change can be estimated based on linear or quadratic margin sensitivity. With this approach tracing of the PV curve for each parameter change can be avoided. This leads to fast estimation of margin for changing conditions. But the prominent sources of inaccuracy inherently associated with margin sensitivity methods make a significant impact on the reliability of the margin estimation. In essence, linear (or high order practically limited to no more than quadratic) sensitivity information is obtained by Taylor series expansion at the system margin point (critical point). Notice that the parameter change, sometimes due to contingency, may not be within a small range and hence higher nonlinearity could not be neglected [2]. Secondly, the effect of system limits may lead to discontinuous change in margin.

This chapter provides methodologies that can be used to estimate the margin for larger change in parameter values.

5.2 Natural Parameterization for Margin Boundary Tracing

As discussed in Chapter 3, power systems can be represented as a Differential and Algebraic Equation (DAE) model and is repeated here.

$$\begin{cases} \dot{X} = F(X,Y,U,Z) \\ 0 = G(X,Y,U,Z) \end{cases} \tag{5.1}$$

X contains all the system state variables; Y includes the algebraic variables; U is the control vector whereas Z consists of load variation at each bus.

Therefore the equilibrium manifold of power system is defined by [3]

$$\begin{cases} 0 = F(X,Y,U,Z) \\ 0 = G(X,Y,U,Z) \end{cases} \tag{5.2}$$

The solution set of above nonlinear equation system constructs a manifold, which could be parameterized by control parameters and disturbance parameters. Both X and Y indicate the state of the system, so they could be combined as state space. The parameter space is the combination of control parameters U and load parameters Z. There is a natural splitting in parameter space.

Parameter space = control parameter space $\oplus$ load parameter space

5.2.1 Load parameter space

As shown in Chapter 3, based on loading scenario, the loading parameter space could be parameterized by scalar λ to characterize the system loading pattern and the corresponding generation change.

$$\begin{cases} P_{Li} = (1 + K_{Li}\lambda)P_{Li0} \\ Q_{Li} = (1 + K_{Li}\lambda)Q_{Li0} \end{cases} \tag{5.3}$$

$$P_{Gi} = (1 + K_{Gi}\lambda)P_{Gi0} \tag{5.4}$$

As mentioned in Chapter 3, K_{Gi} is the generator load picking-up factor that could be determined by AGC, EDC or other system operating practice.

5.2.1 Control parameter space

Control parameter space can contain any type of control of interest. The following controls are studied in this chapter to demonstrate the concepts.

- Shedding   loads
- Shunt capacitance
- Generator secondly voltage control

Control parameter space is parameterized by scalar β to characterize this space

$$U_i = U_{i0} + \beta K_{Ci} \tag{5.5}$$

Where U_{i0} indicates the initial configuration of control i.

Different combinations of control action can be achieved by assigning different ratio value to K_{Ci}.

This parameterization leads to two parameter variations: λ characterizing system loading condition with respect to a specified loading scenario and β characterizing control parameter with respect to a specified control scenario. The equations of power system are reduced to

$$\begin{cases} 0 = F(X, Y, \lambda, \beta) \\ 0 = G(X, Y, \lambda, \beta) \end{cases} \tag{5.6}$$

5.3 Formulation of Margin Boundary Tracing

5.3.1 Margin boundary manifold of power system

In the case of a multi-dimensional, implicitly defined manifold M, specific local parameterization needs to be constructed to trace a certain sub-manifold with special property on M. Saddle node or Hopf bifurcation point forms a margin boundary sub-manifold corresponding to the change of control parameters along a specified control scenario. Therefore bifurcation related stability margin boundary manifold could be traced by augmentation of power system equilibrium with characterization equation.

5.3.2 Characterization of margin boundary

5.3.2.1 Characterization of saddle node bifurcation related margin boundary tracing

Saddle node bifurcation of a dynamical system corresponds to co-dimension 1 fold bifurcation. As discussed in Chapter 3, a cut function for Saddle node related fold bifurcation can be implicitly defined as γ_{SNB} in the following equation

$$\begin{bmatrix} F_X & F_Y & e_k \\ G_X & G_Y & \\ e_j^T & & 0 \end{bmatrix} \begin{bmatrix} u_X \\ u_Y \\ \gamma_{SNB} \end{bmatrix} + \begin{bmatrix} 0 \\ 0 \\ 1 \end{bmatrix} = 0 \qquad (5.7)$$

where we denote $u = \begin{bmatrix} u_X \\ u_Y \end{bmatrix}$, or equivalently

$$\begin{bmatrix} F_X & F_Y & e_k \\ G_X & G_Y & \\ e_j^T & & 0 \end{bmatrix}^T \begin{bmatrix} v_X \\ v_Y \\ \gamma_{SNB} \end{bmatrix} + \begin{bmatrix} 0 \\ 0 \\ 1 \end{bmatrix} = 0 \qquad (5.8)$$

where we denote $v = \begin{bmatrix} v_X \\ v_Y \end{bmatrix}$.

With   the   formulation   of   (5.7),   the   cut   set   condition $\gamma_{SNB}(X,Y,u) = 0$ implies it is at the fold point.

If $\gamma_{SNB}(X,Y,u) = 0$, then

$$\begin{bmatrix} F_X & F_Y \\ G_X & G_Y \end{bmatrix}\begin{bmatrix} u_X \\ u_Y \end{bmatrix} = 0 \quad \text{and} \quad e_j^T\begin{bmatrix} u_X \\ u_Y \end{bmatrix} = 1$$

which implies $\begin{bmatrix} u_X \\ u_Y \end{bmatrix} \neq 0$ and $\begin{bmatrix} F_X & F_Y \\ G_X & G_Y \end{bmatrix}$ is singular.

For this condition $\begin{bmatrix} u_X \\ u_Y \end{bmatrix}$ is the right eigenvector associated with zero ei-

genvalue. Similarly $\begin{bmatrix} v_X \\ v_Y \end{bmatrix}$ is the left eigenvector associated with zero ei-

genvalue.

In principle, the indices k and j in (Eqs.5.7 and 5.8) may be kept fixed throughout the computation, but it is usually advantageous to update them occasionally by selecting new indices for the next step according to (Eqs.5.9 and 5.10)

$$\left|(e^k)^T v \right| = \max\left\{(e^i)^T v \mid, i = 1, \cdots, m\right\} \tag{5.9}$$

$$\left|(e^j)^T u \right| = \max\left\{(e^i)^T u \mid, i = 1, \cdots, m\right\} \tag{5.10}$$

5.3.3 Margin boundary tracing

5.3.3.1 Augmentation for bifurcation characterization

A characterization of bifurcation can be formulated in the cut set form on the solution manifold [4]. Our aim is trace the solution that is on the fold manifold. Solving equation (5.11) with condition (5.12) implies the solution of (5.11) is fold point.

We have to trace the solution of (5.11) for changing values of the load parameter λ and control parameter β.   It becomes a two parameter variation problem

$$B(X,Y,u,\lambda,\beta) = \begin{bmatrix} F(X,Y,\lambda,\beta) \\ G(X,Y,\lambda,\beta) \\ \gamma_{SNB}(X,Y,u) \end{bmatrix} = 0 \qquad (5.11)$$

$$\gamma_{SNB}(X,Y,u) = \begin{bmatrix} F_X & F_Y \\ G_X & G_Y \end{bmatrix} \begin{bmatrix} u_X \\ u_Y \end{bmatrix} = 0 \qquad (5.12)$$

The following sections provide the basic approach to solve these set of equations using predictor and corrector continuation approach we discussed in chapter 3.

5.3.3.2 Augmentation for local parameterization

The total augmented equations for margin boundary tracing are

$$H(X,Y,u,\lambda,\beta) = \begin{bmatrix} B(X,Y,u,\lambda,\beta) \\ \begin{bmatrix} X^T & Y^T & \mu & \lambda & \beta \end{bmatrix} e_k - \eta \end{bmatrix} = [0] \qquad (5.13)$$

$$= \begin{bmatrix} F(X,Y,u,\lambda,\beta) \\ G(X,Y,u,\lambda,\beta) \\ \gamma_{SNB}(X,Y,u) \\ \begin{bmatrix} X^T & Y^T & \mu & \lambda & \beta \end{bmatrix} e_k - \eta \end{bmatrix} = [0]$$

We can solve $H(X,Y,u,\lambda,\beta) = 0$ by applying predictor and corrector approach to (5.13) :

The margin boundary predictor is:

$$\frac{\partial H(X,Y,u,\lambda,\beta)}{\partial(X,Y,u,\lambda,\beta)} \begin{bmatrix} dX \\ dY \\ du \\ d\lambda \\ d\beta \end{bmatrix} = \begin{bmatrix} 0 \\ 0 \\ 0 \\ 0 \\ \pm 1 \end{bmatrix} \tag{5.14}$$

where

$$\frac{\partial H(X,Y,u,\lambda,\beta,)}{\partial(X,Y,u,\lambda,\beta,)} = \begin{bmatrix} \dfrac{\partial F}{\partial X} & \dfrac{\partial F}{\partial Y} & \dfrac{\partial F}{\partial u} & \dfrac{\partial F}{\partial \lambda} & \dfrac{\partial F}{\partial \beta} \\ \dfrac{\partial G}{\partial X} & \dfrac{\partial G}{\partial Y} & \dfrac{\partial G}{\partial u} & \dfrac{\partial G}{\partial \lambda} & \dfrac{\partial G}{\partial \beta} \\ \dfrac{\partial C}{\partial X} & \dfrac{\partial C}{\partial Y} & \dfrac{\partial C}{\partial u} & \dfrac{\partial C}{\partial \lambda} & \dfrac{\partial C}{\partial \beta} \\ & & e_k & & \end{bmatrix} \tag{5.15}$$

After solving (5.14) for a tangent vector, the predicted values of the unknown variables can be obtained from (5.15). Where δ is the step length

$$\begin{bmatrix} X_{pre} \\ Y_{pre} \\ u_{pre} \\ \lambda_{pre} \\ \beta_{pre} \end{bmatrix} = \begin{bmatrix} X \\ Y \\ u \\ \lambda \\ \beta \end{bmatrix}^{[i]} + \delta \begin{bmatrix} dX \\ dY \\ du \\ d\lambda \\ d\beta \end{bmatrix} \tag{5.16}$$

This predicted value can be used to as an initial guess to converge upon the margin boundary by solving the non-linear algebraic equations (5.13) with the Newton-Raphson method

Boundary corrector

Newton method is employed to do the boundary correction as

$$
\begin{bmatrix} X \\ Y \\ u \\ \lambda \\ \beta \end{bmatrix}^{new} = \begin{bmatrix} X \\ Y \\ u \\ \lambda \\ \beta \end{bmatrix} - (\frac{\partial H(X,Y,u,\lambda,\beta)}{\partial(X,Y,u,\lambda,\beta)})^{-1} H(X,Y,u,\lambda,\beta) \tag{5.17}
$$

Iterate until the mismatch is less than the tolerance. Finally one can obtain the solution which is the fold point corresponding to $\lfloor X^T, Y^T, u, \lambda, \beta \rfloor e_k = \eta$

5.3.4 Basic Steps Involved in the Margin Boundary Tracing

The following steps are involved in margin boundary tracing.

1. Specify a loading scenario.
2. Equilibrium Tracing Program (EQTP) starts at current operating point for the first boundary point under current fixed control configuration and specified loading scenario.
3. Specify the control scenario that describes the change of control configuration or contingencies.
4. Predict the Boundary with Eq.5.16.
5. Correct the Boundary with Eq.5.17.
6. Go to step 4 unless some control variables hit limits, else stop.

5.3.5 Practical implementation

In the previous section saddle node bifurcation condition is explicitly included in the set of nonlinear equations. So when you solve these equations for changing load and control parameters the solution is always on the boundary. This formulation needs the second order derivatives. Another way to trace these boundaries is by extending EQTP discussed in Chapter 3. This approach is briefly explained through Fig.5.1.

For practical control variables range one may not encounter co-dimension 2 bifurcation (where the rank of the system Jacobian is n-2). In that case, a reduced method with only a one-dimension augmentation (unfolding) can be easily employed to effectively trace voltage stability margin boundary. Nevertheless, since only one-dimension augmentation (unfolding) is applied, theoretically the reduced tracing method has a limited tracing range and could diverge near a co-dimension 2 saddle node bifurcation.

5.3.5.1 Implementation of reduced method

As for the power system equilibrium manifold in Eq.5.6, at first we fix the control at base value (β_0). Then

$$\begin{cases} 0 = F(X,Y,\lambda,\beta_0) \\ 0 = G(X,Y,\lambda,\beta_0) \end{cases}$$

When the system is not at a neighborhood of a co-dimension 2 saddle node bifurcation, the e_k in the second augmentation in Eq.5.13 could be set so that it always select λ as the continuation parameter. Then the problem turns into solving equation F, G, and cut function γ_{SNB} under different specified control conditions characterized by λ.

Notice that the previous margin boundary points are used as pivot (cut) condition to calculate the initial points of the part of equilibrium trajectory leading to the next margin boundary point defined by the new control parameter. Still the system load margin corresponding to a new control is determined without retracing the entire PV curve. Therefore the reduced margin boundary tracing is also computationally efficient. The procedure is as follows.

1. Under base control β_0 , use EQTP [5] and bifurcation identification conditions to trace to the SNB and then the initial margin boundary point is obtained.

$$\text{Predictor:} \begin{bmatrix} F_X & F_Y & F_\lambda \\ G_X & G_Y & G_\lambda \\ & e_j^T & \end{bmatrix}^{\beta_0} \begin{bmatrix} dX \\ dY \\ d\lambda \end{bmatrix} = \begin{bmatrix} 0 \\ 0 \\ \pm 1 \end{bmatrix} \tag{5.18}$$

$$\text{Corrector:} \begin{bmatrix} F_X & F_Y & F_\lambda \\ G_X & G_Y & G_\lambda \\ & e_j^T & \end{bmatrix}^{\beta_0} \begin{bmatrix} \Delta X \\ \Delta Y \\ \Delta\lambda \end{bmatrix} = - \begin{bmatrix} F^{\beta_0} \\ G^{\beta_0} \\ 0 \end{bmatrix} \tag{5.19}$$

Assume that the initial margin boundary point with load margin λ_0 under the base control β_0 is obtained.

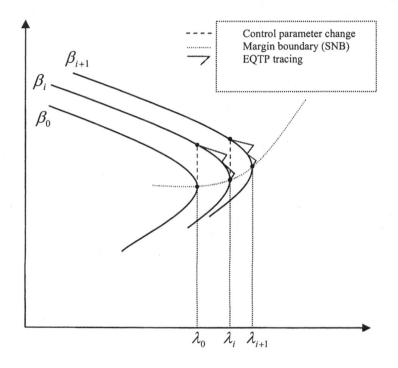

Fig. 5.1 Illustration of reduced implementation of margin boundary Tracing

2. Change the control parameter by a specified single step size δ. $\beta_{i+1} = \beta_i + \delta$.

3. With the jth element of the previous margin boundary point solution vector $[X^T \ Y^T \ \lambda_0]^T$ as the pivot (cut) condition, solve the following equations and use the solution as the initial point to trace the new equilibrium trajectory defined by the new control parameter.

4. From the new initial point, use EQTP and bifurcation identification conditions to trace the next SNB under the new control parameter β_{i+1}.

$$\text{Predictor:} \quad \begin{bmatrix} F_X & F_Y & F_\lambda \\ G_X & G_Y & G_\lambda \\ & e_j^T & \end{bmatrix}^{\beta_{i+1}} \begin{bmatrix} dX \\ dY \\ d\lambda \end{bmatrix} = \begin{bmatrix} 0 \\ 0 \\ \pm 1 \end{bmatrix} \tag{5.20}$$

$$\text{Corrector:} \quad \begin{bmatrix} F_X & F_Y & F_\lambda \\ G_X & G_Y & G_\lambda \\ & e_j^T & \end{bmatrix}^{\beta_{i+1}} \begin{bmatrix} \Delta X \\ \Delta Y \\ \Delta \lambda \end{bmatrix} = - \begin{bmatrix} F^{\beta_{i+1}} \\ G^{\beta_{i+1}} \\ 0 \end{bmatrix} \tag{5.21}$$

Then a new boundary point with margin λ_{i+1} under the new control parameter β_{i+1} is obtained (Note: the middle curve in the Fig.5.1 corresponds to the control β_i).

5. Repeat step 2-4 to obtain the margin boundary points until the studied controls hit their limits.

5.4 Examples

Two bus example:

For this example, the unity power factor is still used. Margin boundary tracing can be demonstrated with respect to shunt compensation and series compensation, respectively. The method described in section 5.3.3 is first utilized here to trace the margin boundary. Then, the method described in section 5.3.4 is applied for comparison. For simplification, only power flow equations are used.

(1) Shunt compensation at bus 2

The power flow equations of the 2-bus system are:

$$G_1(\delta, V_2, \underline{h}, \lambda, \beta) = 1.4(1+\lambda) + 10V_2 \sin(\delta) = 0$$
$$G_2(\delta, V_2, \underline{h}, \lambda, \beta) = -10V_2 \cos(\delta) + (10-\beta)V_2^2 = 0$$

where, β is the shunt capacitance at bus 2(p.u.).

Singularity conditions are shown as follows:

$$\begin{bmatrix} G_{1\delta} & G_{1V_2} \\ G_{2\delta} & G_{2V_2} \end{bmatrix}\begin{bmatrix} h_1 \\ h_2 \end{bmatrix} = \begin{bmatrix} 10V_2\cos(\delta) & 10\sin(\delta) \\ 10V_2\sin(\delta) & (20-2\beta)V_2 - 10\cos(\delta) \end{bmatrix}\begin{bmatrix} h_1 \\ h_2 \end{bmatrix} = \begin{bmatrix} 0 \\ 0 \end{bmatrix}$$

$$h_2 = 1.0$$

Since h_2 is a constant, we can replace it with 1.0.

$$1.4(1+\lambda) + 10V_2\sin(\delta) = 0$$
$$-10V_2\cos(\delta) + (10-\beta)V_2^2 = 0$$
$$10V_2\cos(\delta)h_1 + 10\sin(\delta) = 0$$
$$10V_2\sin(\delta)h_1 + [(20-2\beta)V_2 - 10\cos(\delta)] = 0$$

[Note: For this simple 2-bus example, we can also use Eq.5.7 to get the cut function for the bifurcation boundary; it can be used to replace the last two equations of the above set.

$$\gamma(\delta,V_2,\lambda,\beta) = 10 - (20-2\beta)V_2\cos(\delta) = 0\,]$$

Hence, the H matrix can be expressed as:

$$H = \begin{bmatrix} 10V_2\cos(\delta) & 10\sin(\delta) & 0 & 1.4 & 0 \\ 10V_2\sin(\delta) & (20-2\beta)V_2 - 10\cos(\delta) & 0 & 0 & -V_2^2 \\ -10V_2\sin(\delta)h_1 + 10\cos(\delta) & 10\cos(\delta)h_1 & 10V_2\cos(\delta) & 0 & 0 \\ 10V_2\cos(\delta)h_1 + 10\sin(\delta) & 10\sin(\delta)h_1 + 20 - 2\beta & 10V_2\sin(\delta) & 0 & -2V_2 \\ & e_k^T & & & \end{bmatrix}$$

Select β as the continuation parameter, then $k=5$.

Tangent vector is: $[d\delta \quad dV_2 \quad dh_1 \quad d\lambda \quad d\beta]^T$

Base case SNB point is (corresponding to $\beta = 0$, $P_0 = 0.14$, normalized value):

$$V_2 = 0.7071, \delta = -0.7854, \lambda = 2.5714, h_1 = 1.4142, P_{cri} = 0.5$$

Predictor step:

$$[d\delta \quad dV_2 \quad dh_1 \quad d\lambda \quad d\beta]^T$$
$$= [0.0079 \quad 0.0763 \quad -0.1302 \quad 0.3571 \quad 1.0]^T$$

Taking step length $\sigma = 0.1$, then:

$$[\delta^{pre} \quad V_2^{pre} \quad h_1^{pre} \quad \lambda^{pre} \quad \beta^{pre}]^T$$
$$= [-0.7846 \quad 0.7147 \quad 1.4012 \quad 2.6071 \quad 0.1]^T$$

Corrector step:

$$[\Delta\delta \quad \Delta V_2 \quad \Delta h_1 \quad \Delta\lambda \quad \Delta\beta]^T$$
$$= [-0.0008 \quad -0.0005 \quad -0.0012 \quad 0.0004 \quad 0]^T$$

$$[\delta^{new} \quad V_2^{new} \quad h_1^{new} \quad \lambda^{new} \quad \beta^{new}]^T$$
$$= [-0.7854 \quad 0.7142 \quad 1.40 \quad 2.6075 \quad 0.1]^T$$

Hence, the new margin is:

$$P_0 * (1 + \lambda^{new}) = 0.14 * (1 + 2.6075) = 0.5051$$

Applying the continuation Prediction-Correction method step by step, we can trace the saddle node bifurcation margin boundary with respect to β (shunt capacitance).

Furthermore, if we use the normalized equation, the analytical expression for the margin for this simple system is given in the reference [6]:

$$p^{new} = \frac{\cos(\varphi)}{1 + \sin(\varphi)} \frac{1}{1 - B_c X}$$

Where, B_c is the shunt capacitance at bus 2(p.u.).

The comparison results between analytical value and MBT approach with the intermediate steps are shown in table5.1 and fig5.2:

Table 5.1 Comparison results

Shunt2 (p.u.)	V_2 (p.u.)	h_1	λ	MBT (normz.)	Analytical (normz.)
0	0.7071	1.4142	2.5714	0.5	0.5
0.1	0.7142	1.4001	2.6075	0.505	0.50505
0.2	0.7215	1.3859	2.6443	0.5102	0.5102
0.3	0.729	1.3718	2.6819	0.5155	0.51546
0.4	0.7366	1.3576	2.7202	0.5208	0.52083
0.5	0.7443	1.3435	2.7594	0.5263	0.52632
0.6	0.7522	1.3294	2.7994	0.5319	0.53191
0.7	0.7603	1.3152	2.8402	0.5376	0.53763
0.8	0.7686	1.3011	2.882	0.5435	0.54348
0.9	0.777	1.2869	2.9246	0.5495	0.54945
1	0.7857	1.2728	2.9683	0.5556	0.55556
1.1	0.7945	1.2587	3.0128	0.5618	0.5618
1.2	0.8035	1.2445	3.0584	0.5682	0.56818
1.3	0.8128	1.2304	3.1051	0.5747	0.57471
1.4	0.8222	1.2162	3.1528	0.5814	0.5814
1.5	0.8319	1.2021	3.2017	0.5882	0.58824

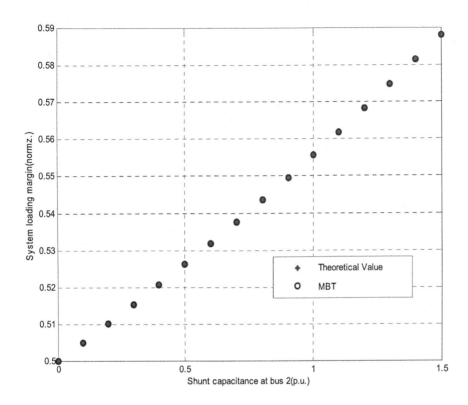

Fig. 5.2 System margin variation with shunt capacitance (analytical vs. MBT full)

Figure 5.3 shows the margin variation obtained through reduced formulation. These results are comparable.

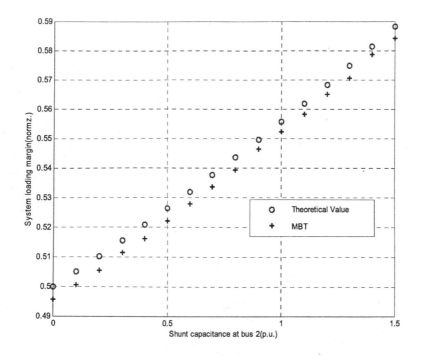

Fig. 5.3 System margin variation with shunt capacitance (analytical vs.

MBT reduced)

(2) Series compensation between bus 1 and bus 2

The formulation is same as before except here β is equivalent total se-ries admittance

$$G_1(\delta, V_2, \lambda, \beta) = 1.4(1 + \lambda) + \beta V_2 \sin(\delta) = 0$$
$$G_2(\delta, V_2, \lambda, \beta) = -\beta V_2 \cos(\delta) + \beta V_2^2 = 0$$

Singularity conditions:

$$\begin{bmatrix} G_{1\delta} & G_{1V_2} \\ G_{2\delta} & G_{2V_2} \end{bmatrix} \begin{bmatrix} h_1 \\ h_2 \end{bmatrix} = \begin{bmatrix} \beta V_2 \cos(\delta) & \beta \sin(\delta) \\ \beta V_2 \sin(\delta) & 2\beta V_2 - \beta \cos(\delta) \end{bmatrix} \begin{bmatrix} h_1 \\ h_2 \end{bmatrix} = \begin{bmatrix} 0 \\ 0 \end{bmatrix}$$
$$h_2 = 1.0$$

The total set of equations to be solved:

$$1.4(1+\lambda)+\beta V_2 \sin(\delta)=0$$
$$-\beta V_2 \cos(\delta)+\beta V_2^2=0$$
$$\beta V_2 \cos(\delta)h_1+\beta \sin(\delta)=0$$
$$\beta V_2 \sin(\delta)h_1+2\beta V_2-\beta \cos(\delta)=0$$

$$H=\begin{bmatrix} \beta V_2 \cos(\delta) & \beta \sin(\delta) & 0 & 1.4 & V_2 \sin(\delta) \\ \beta V_2 \sin(\delta) & 2\beta V_2-\beta \cos(\delta) & 0 & 0 & V_2^2-V_2 \cos(\delta) \\ -\beta V_2 \sin(\delta)h_1+\beta \cos(\delta) & \beta \cos(\delta)h_1 & \beta V_2 \cos(\delta) & 0 & V_2 \cos(\delta)h_1+\sin(\delta) \\ \beta V_2 \cos(\delta)h_1+\beta \sin(\delta) & \beta \sin(\delta)h_1+2\beta & \beta V_2 \sin(\delta) & 0 & V_2 \sin(\delta)h_1+2V_2-\cos(\delta) \\ & e_k^T & & & \end{bmatrix}$$

Select β as the continuation parameter, then $k=5$.

Tangent vector is: $[d\delta \quad dV_2 \quad dh_1 \quad d\lambda \quad d\beta]^T$.

Base case SNB point is (corresponding to $\beta=10$, and $P_0=0.14$ is normalized value):

$$V_2=0.7071, \delta=-0.7854, \lambda=2.5714, h_1=1.4142, P_{cri}=0.5$$

Applying the continuation Prediction-Correction method step by step as shown before one can trace the saddle node bifurcation margin boundary with respect to β (series admittance).

The analytical expression of the margin for series compensation as given in reference [6] is:

$$P_{max}=\frac{\cos(\varphi)}{1+\sin(\varphi)}\frac{E^2}{2X}=\frac{\cos(\varphi)}{2(1+\sin(\varphi))}E^2B$$

This relation shows that the margin changes linearly with respect to the change of equivalent total series admittance.

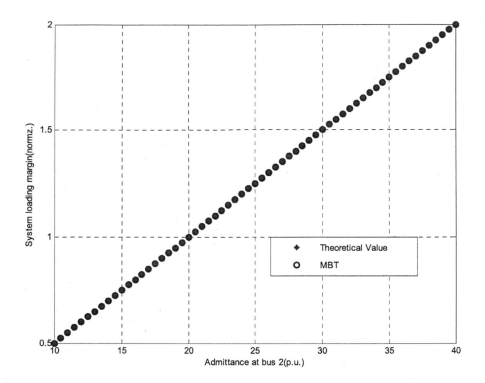

Fig.5.4   System margin variation with total equivalent series admittance

(analytical vs. MBT)

Meanwhile, the degree of compensation can be calculated by:

$$X_c\% = (1 - \frac{B_0}{B_{new}}) * 100$$

If we use the $X_c\%$ as the parameter β instead, the power flow equations will be:

$$G_1(\delta, V_2, \lambda, \beta) = 1.4(1 + \lambda) + \frac{1000}{100 - \beta} V_2 \sin(\delta) = 0$$

$$G_2(\delta, V_2, \lambda, \beta) = -\frac{1000}{100 - \beta} V_2 \cos(\delta) + \frac{1000}{100 - \beta} V_2^2 = 0$$

And H changes into (using $m = \dfrac{1000}{100-\beta}, n = \dfrac{1000}{(100-\beta)^2}$ to simplify):

$$H = \begin{bmatrix} mV_2\cos(\delta) & m\sin(\delta) & 0 & 1.4 & nV_2\sin(\delta) \\ mV_2\sin(\delta) & mV_2\cos(\delta) & 0 & 0 & n(V_2^2 - V_2\cos(\delta)) \\ -mV_2h_1\sin(\delta)+m\cos(\delta) & m\cos(\delta)h_1 & mV_2\cos(\delta) & 0 & n[V_2\cos(\delta)h_1+\sin(\delta)] \\ mV_2\cos(\delta)h_1 + m\sin(\delta) & m\sin(\delta)h_1+2m & mV_2\sin(\delta) & 0 & n[V_2\sin(\delta)h_1+2V_2-\cos(\delta)] \\ & & e_k^T & & \end{bmatrix}$$

The figure5.5 shows the change of system loading margin with respect to degree of compensation $X_c\%$:

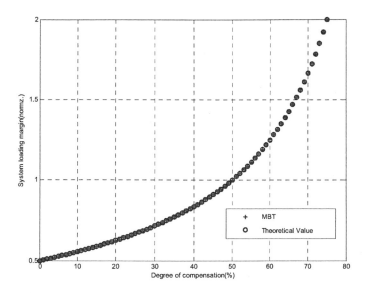

Fig.5.5 System margin variation with degree of compensation (analytical vs. MBT)

New England 39-bus system:

The following assumptions are made to demonstrate the boundary tracing on this test system

• Constant power load model;

- The maximum real power limit, the field current and the armature current limits are considered for each generator.
- No generator is allowed to have terminal voltage higher than 1.1 p.u., when its secondary voltage control is utilized to increase system stability margin;
- The loading scenario is defined as that all the loads are increased wit constant power factor, and all the generators participate in the load pickup at the same rate.

The margin boundaries can be traced with respect to any specified control scenario.

5.4.1 Series compensation between bus 6 and bus 31

The figure5.6 shows the system loading margin change as the series admittance between bus 6 and bus 31 varies.

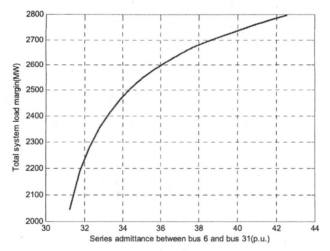

Fig. 5.6 System loading margin vs. series compensation

5.4.2 Shunt Compensation

Fig.5.7 shows the system loading margin change as shunt capacitance increases at bus 10.

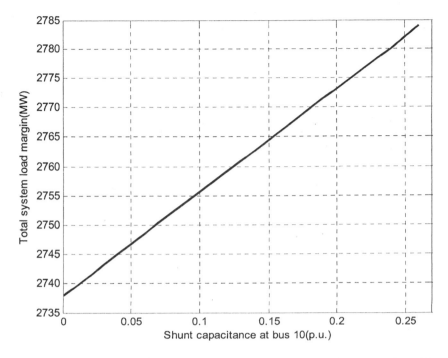

Fig. 5.7 System load margin vs. shunt capacitance at bus 10

5.4.3 Multiple contingencies

Voltage stability margin change due to single or multiple contingencies could also be traced by parameterizing the control parameter change involved in the contingency.

Let the $Y_{ij} = Y_{ij}^{(0)}(1-\beta)$ and $B_{ij} = B_{ij}^{(0)}(1-\beta)$. When the parameter β varying from zero to one, Y_{ij} and B_{ij} will vary from their initial values of $Y_{ij}^{(0)}$ and $B_{ij}^{(0)}$ to zero. Therefore, Y matrix becomes the post-contingency value.

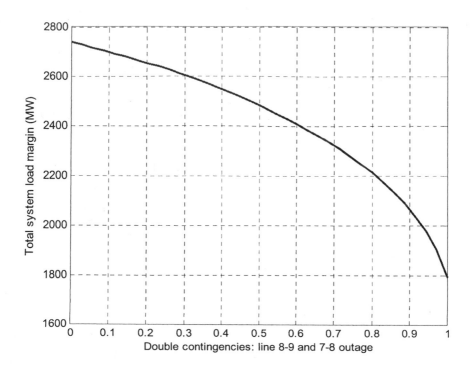

Fig. 5.8 System load margin vs. multiple contingencies:
line 8-9 and line 7-8 outages

Fig. 5.8 shows the margin change for line outages. Zero indicates both line
are in and one indicates both lines are out.

5.4.4 Boundary tracing with respect to generation control parameters

The margin boundaries can be traced with respect to any specified control
scenario.

5.4.4.1 Load margin vs adjustment of Ka of AVR system

In Fig.5.9, voltage collapse (SNB related) margin boundary versus adjustment of Ka around its base case operating value is depicted as the solid
curve.

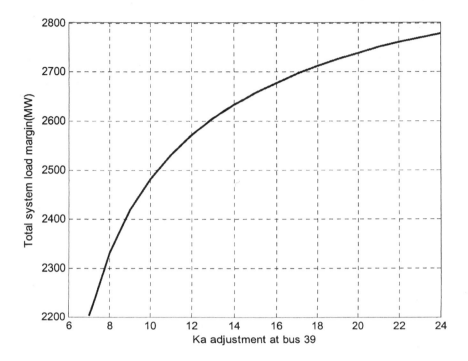

Fig. 5.9 margin boundary tracing vs. Ka adjustment

5.4.4.2 Load margin versus adjustment of V_{ref} of AVR system

Fig.5.10 shows the system voltage stability margin change with respect to the change of generator V_{ref} at bus 39.

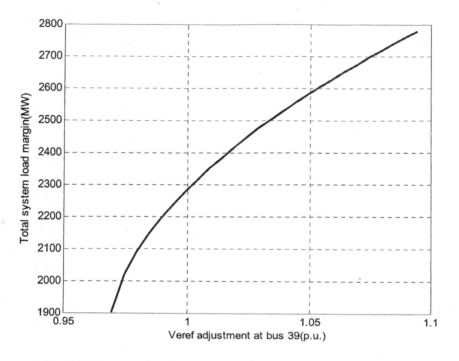

Fig. 5.10 System load margin vs. V_{ref} adjustment at generator bus 39

5.4.5 Control combination

The control scenario could be any combination of control parameters. Fig. 5.11 shows the variation of margin with simultaneous change in V_{ref}, and shunt compensation(The V_{ref} is increased from 1.084 p.u. to 1.092 p.u., and shunt capacitance is increased from 0 to 0.08 p.u.).

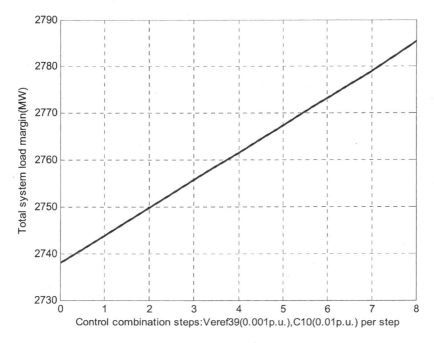

Fig. 5.11 System load margin vs. control combination steps: Vref39 (0.001pu), C10 (0.01pu)

5.4.6 Advantages of margin boundary tracing

- Margin Boundary Tracing is accurate and reliable.
- It is easy to take account of limit effects and other nonlinearities in Mar gin Boundary Tracing.
- Margin boundary tracing dramatically saves CPU time compare to ob taining each new boundary point by exhaustively recomputing the whole PV curve.

5.5 Formulation of Voltage Stability Limited ATC

Deregulation in power industries is promoting the open access of all transmission networks. However, this may lead to the violation of transfer capability in the networks. These aspects have motivated the development of methodologies to evaluate existing power transfer capabilities and transmission margins. The term "Available Transfer Capability" (ATC) [7] is used to measure the transfer capability remaining in the physical transmission network for further commercial activity over already committed uses.

A key aspect in calculating ATC is the physical and operational limitations [7, 8] of the transmission system, such as circuit ratings and bus voltage levels. In addition, as power system become more heavily loaded, voltage collapse is more likely to occur.

To determine ATC is actually to determine TTC (Total Transfer Capability), which is the most critical physical or operational limit to the networks. TTC on some portions of the transmission network shifts among thermal, voltage and stability limits as the network operating conditions change over time. Fig.5.12 shows one possible ATC scenario with the low voltage limit is the critical constraint while in Fig.5.13 the voltage stability limit is the critical constraint [9].

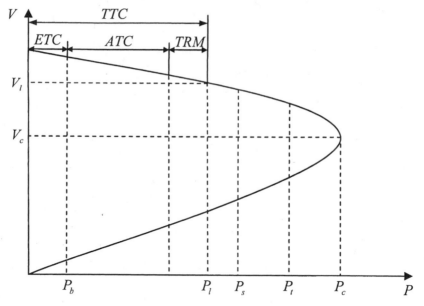

Fig.5.12 Illustration of ATC with low voltage limit as the critical constraint

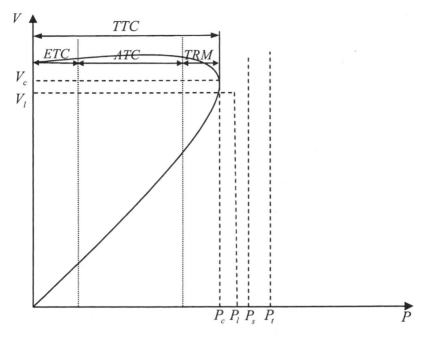

Fig.5.13 Illustration of ATC with voltage stability limit as the critical constraint

- V_l : Low voltage with respect to bus voltage limit;
- V_c : Critical voltage with respect to voltage collapse point;
- P_b : Existing Transmission Commitments (including CBM);
- P_l : Low voltage limit;
- P_s : Oscillatory stability or transient stability limit;
- P_t : Thermal overload limit;
- P_c : Voltage stability limit;
- TTC : $\min(P_l, P_s, P_t, P_c)$

The continuation power flow and the margin boundary tracing techniques can be applied to obtain voltage stability limited transactions.

For this we have to obtain scenario parameters that establish various transactions including multi-area simultaneous transactions. Next section describes how to obtain scenario parameters to establish a particular transaction.

5.6 Scenario Parameters

Recall equilibrium tracing introduced in chapter 3 is along a fixed load/generation changing pattern from the base case up to voltage collapse point. The load/generation changing pattern at each bus in the system is reproduced here for the continuity.

$$P_{Li}(\lambda) = P_{Li0} + \lambda K_{Li} P_{Li0} \tag{5.22}$$

$$Q_{Li}(\lambda) = Q_{Li0} + \lambda K_{Li} Q_{Li0} \tag{5.23}$$

$$P_{Gi}(\lambda) = P_{Gi0} + K_{Gi} \sum_{i=1}^{N} (P_{Li}(\lambda) - P_{Li0}) \tag{5.24}$$

$$\sum_{i=1}^{N} K_{Gi} = 1$$

Note that K_L s are not independent variables. They do not have any meaning if they are not compared to each other since K_{Li} only reflects the relative load change speed at one specific bus. For instance, if a system has only two buses where the load will increase, the case with $K_{L1} = 1$ and $K_{L2} = 2$ equals to the case where $K_{L1} = 2$ and $K_{L2} = 4$. In other words, it is the ratio between K_L s that determines the system active load increase scenario.

K_{Gi} reflects active generation dispatching pattern at each bus. When there is no generation increase at a particular bus i, the corresponding K_{Gi} is zero.

In summary, when the common λ increases, the active load, reactive load, and generation at each bus will increase at the rate determined by K_L s, and K_G s, respectively.

Keeping the essential meaning of K_L s, and K_G s in mind, we will see how they are extended in the following section to simulate the process of simultaneous multi-area transactions.

5.7 Scenario According to Simultaneous Multi-area Transactions

As shown in Fig.5.14, there is more than one desired transaction between n areas, which will take place at the same time. Since the n areas are inter-connected, their performance will affect one another.

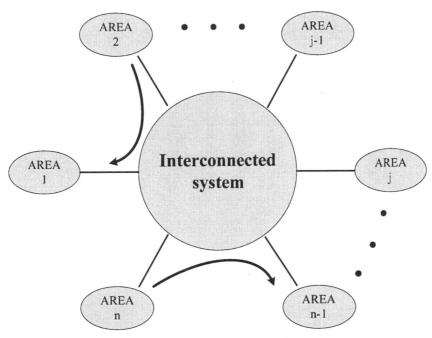

Fig.5.14 Multi-area transactions

In addition, one area may have several transactions with others at the same time either as seller or as buyer. For a highly meshed system, an explicit interface may not exist between areas. Thus choosing the interface flow to check whether these transactions are satisfied may not be reasonable. Alternatively, we can set up a criterion for judgment by checking the seller's supply and buyer's load increase. That is, according to the transactions associated to each area (either seller or buyer), we can summarize the total demand for each seller and requirement of each buyer. If at one point, each seller reaches its total supply demand while at the same time each buyer reaches its total requirement, we claim these transactions are satisfied simultaneously no matter how the flows go through the network. This criterion is also more suitable for studying voltage-stability-related problems since voltage stability is inherently related to system's load/generation pattern, it cannot simply be judged by the flow over the transmission line like what is done to study thermal constraint.

Given the demand of each transaction between n areas (generalized in Table 5.2), we can obtain the total desired load demand for each area as well as the total desired generation supply for each area according to the transactions, which is shown in Table 5.3. For the sake of generality, we also consider transactions within the same area, e.g., $i=j$. Table 5.4 shows the total number of load increase buses and the total number of generation-sharing buses of each area.

Table 5.2 Transaction description

Area # of seller	Area # of buyer	Desired transaction amount (MW)
i	$j(j \neq i)$	P_{Tij}
i	i	P_{Tii}

Table 5.3 Demand for each area's generation and load

Area #	Demand for load (MW)	Demand for generation (MW)
1	$\sum_{i=1}^{n} P_{Ti1}$	$\sum_{i=1}^{n} P_{T1i}$
$\vdots$	$\vdots$	$\vdots$
j	$\sum_{i=1}^{n} P_{Tij}$	$\sum_{i=1}^{n} P_{Tji}$
$\vdots$	$\vdots$	$\vdots$
n	$\sum_{i=1}^{n} P_{Tin}$	$\sum_{i=1}^{n} P_{Tni}$

According to the information above, we need to determine the load parameters K_Ls and generation parameters K_Gs to simulate these transactions. For one-area case, scenario parameters can be set only according to the demand of load and generation at each bus in this particular area, without considering other area. So the load changing parameter, K_Ls and generation changing parameter, K_Gs can be determined arbitrarily as long as the total generation and load in the area can be balanced. However, for simultaneous multi-area transactions, merely using such strategy may cause a mismatch between the seller's actual generation and the corresponding buyer's demand. Now that each area's scenario is considered within a combined large system, we need to guarantee the simultaneousness as well to match the seller's generation output to the corresponding buyer's load increase.

Table 5.4 Area scenario

Area #	Table # of genera-tion increase buses	Total # of load increase buses
1	NG_1	NL_1
$\vdots$	$\vdots$	$\vdots$
j	NG_j	NL_j
$\vdots$	$\vdots$	$\vdots$
n	NG_n	NL_n

5.7.1 Determination of K_{Li}

According to the definition of simultaneous transactions, each area will increase its load in such a way that when it satisfies its own load demand according to transactions, other areas will also exactly satisfy their own load demand according to the transactions requirement. Without the loss of generality, we can assume that all involved areas increase their loads in a simultaneous way. That is, we can determine the load increase scenario according to the ratio of each area's load demand with respect to total load demand for the whole equilibria-tracing process from the base case, which guarantees the simultaneousness of a set of transactions.

Note that the load demand for each area (in Table 5.4) can be regarded as the net load increase from the base case for that particular area corresponding to the bifurcation parameter at that point, λ. Since we have defined the simultaneousness of the set of transactions, λ is common for each area, which can be mathematically expressed with the following equation set (compared to Eq.5.22):

$$\lambda \sum_{k=1}^{NL_1} K_{Lk}^{(1)} P_{Lk0}^{(1)} = \sum_{i=1}^{n} P_{Ti1}$$

$$\vdots$$

$$\lambda \sum_{k=1}^{NL_j} K_{Lk}^{(j)} P_{Lk0}^{(j)} = \sum_{i=1}^{n} P_{Tij}$$

$$\vdots$$

$$\lambda \sum_{k=1}^{NL_n} K_{Lk}^{(n)} P_{Lk0}^{(n)} = \sum_{i=1}^{n} P_{Tin} \qquad (5.25)$$

where P_{Lk0} refers to the existing load at bus k and the upper note (j) specifies the area in which that bus resides.

By summarizing the left-hand-sides and right-hand sides of each equation in equation set 5.25, respectively, we have

$$\lambda \sum_{j=1}^{n} \sum_{k=1}^{NL_j} K_{Lk}^{(j)} P_{Lk0}^{(j)} = \sum_{j=1}^{n} \sum_{i=1}^{n} P_{Tij} \qquad (5.26)$$

The right-hand-side of Eq.5.26 is nothing but the total transactions' demand. We simply denote it as follows:

$$P_{TOTAL} = \sum_{j=1}^{n} \sum_{i=1}^{n} P_{Tij} \qquad (5.27)$$

Taking one equation related to AREA j from Eq.5.25 and dividing it by Eq.5.26, we get

$$\frac{\sum_{k=1}^{NL_j} K_{Lk}^{(j)} P_{Lk0}^{(j)}}{\sum_{j=1}^{n} \sum_{k=1}^{NL_j} K_{Lk}^{(j)} P_{Lk0}^{(j)}} = \frac{\sum_{i=1}^{n} P_{Tij}}{P_{TOTAL}} \qquad (5.28)$$

We normalize K_Ls by letting

$$\sum_{j=1}^{n} \sum_{k=1}^{NL_j} K_{Lk}^{(j)} P_{Lk0}^{(j)} = C \qquad (5.29)$$

where C is an arbitrary constant in MW. Then Eq.5.28 is transformed to

$$\sum_{k=1}^{NL_j} K_{Lk}^{(j)} P_{Lk0}^{(j)} = C \frac{\sum_{i=1}^{n} P_{Tij}}{P_{TOTAL}} \tag{5.30}$$

Eq.5.30 reflects when the total systems' load changes; each area's load changes according to the fraction determined by the transactions.

The purpose of normalizing K_Ls is mainly to keep bifurcation parameter λ and system transfer (in MW) consistent and make these two have simple mathematical relationship. Choosing C is arbitrary because it affects only the value of K_Ls but not their essence (see Section 5.6).

To finally determine K_Ls, we further need the load change relationship between each bus in a particular area, which can be given in the following form:

$$K_{L1}^{(j)} : \cdots : K_{Lk}^{(j)} : \cdots : K_{L,NL_j}^{(j)} = \mu_1^{(j)} : \cdots : \mu_k^{(j)} : \cdots : \mu_{NL_j}^{(j)} \tag{5.31}$$

where $\mu_k^{(j)}$ is a constant coefficient. By introducing a common variable for each area j, $K_{L\_BASE}^{(j)}$, we can transform Eq.5.31 to

$$K_{L1}^{(j)} = \mu_1^{(j)} K_{L\_BASE}^{(j)}$$
$$\vdots$$
$$K_{Lk}^{(j)} = \mu_k^{(j)} K_{L\_BASE}^{(j)}$$
$$\vdots$$
$$K_{L,NL_j}^{(j)} = \mu_{NL_j}^{(j)} K_{L\_BASE}^{(j)} \tag{5.32}$$

Substituting Eq.5.32 into 5.30, we get

$$K_{L\_BASE}^{(j)} = C \frac{1}{\sum_{k=1}^{NL_j} \mu_k^{(j)} P_{Lk0}^{(j)}} \frac{\sum_{i=1}^{n} P_{Tij}}{P_{TOTAL}} \tag{5.33}$$

Substituting Eq.5.33 into 5.32, we finally obtain K_L for each bus in each area as follows:

$$K_{Lk}^{(j)} = C \frac{\mu_k^{(j)}}{\sum_{k=1}^{NL_j} \mu_k^{(j)} P_{Lk0}^{(j)}} \frac{\sum_{i=1}^{n} P_{Tij}}{P_{TOTAL}} \qquad (5.34)$$

Through the above procedure, the obtained K_L s reflect the load change relationship between each bus strictly according to the transactions.

5.7.2 Determination of K_{Gi}

According the principle of simultaneous transactions, each generation area needs to pick up the load demand associated with it. From the whole system point of view, each generation area only picks up part of the total load demand. Thus we can assign the ratio of generation sharing for one particular area according to Table 5.3 as follows:

$$\sum_{k=1}^{NG_j} K_{Gk}^{(j)} = \frac{\sum_{i=1}^{n} P_{Tji}}{P_{TOTAL}} \qquad (5.35)$$

Eq.5.35 shows that each generating area picks up part of total load according to the transaction associated with it. To finally determine $K_G^{(j)}$s, we further need the generation-sharing relationship between each bus in a particular area, which can be given in the following form:

$$K_{G1}^{(j)} : \cdots : K_{Gk}^{(j)} : \cdots : K_{G,GL_j}^{(j)} = \eta_1^{(j)} : \cdots : \eta_k^{(j)} : \cdots : \eta_{GL_j}^{(j)} \qquad (5.36)$$

where $\eta_k^{(j)}$ is a constant coefficient that reflects the relative generation-pick up of one particular bus. By introducing a common variable for each area j, $K_{G\_BASE}^{(j)}$, we can transform Eq.5.36 to

$$K_{G1}^{(j)} = \eta_1^{(j)} K_{G\_BASE}^{(j)}$$

$$\vdots$$

$$K_{Gk}^{(j)} = \eta_k^{(j)} K_{G\_BASE}^{(j)}$$

$$\vdots$$

$$K_{G,NG_j}^{(j)} = \eta_{NG_j}^{(j)} K_{G\_BASE}^{(j)} \qquad (5.37)$$

Substituting Eq.5.37 into 5.35, we get

$$K_{G\_BASE}^{(j)} = \frac{1}{\sum_{k=1}^{NG_j} \eta_k^{(j)}} \frac{\sum_{i=1}^{n} P_{Tji}}{P_{TOTAL}}$$ (5.38)

Substituting Eq.5.38 into 5.37, we finally obtain K_G for each bus in each area as follows:

$$K_{Gk}^{(j)} = \frac{\eta_k^{(j)}}{\sum_{k=1}^{NG_j} \eta_k^{(j)}} \frac{\sum_{i=1}^{n} P_{Tji}}{P_{TOTAL}}$$ (5.39)

Furthermore, this methodology is also applicable to more general cases such as multiple transactions from one area or load and generation increase in the same area.

5.8 Numerical Example

The procedure described in Section 5.3 is implemented in the EQTP simulation to show the effectiveness of direct ATC tracing. Various control scenarios have been traced for voltage stability related ATC margin boundary. Simultaneous multi-area transactions are defined in the simulation to show their impact on ATC margin for different areas.

5.8.1 Description of the simulation system

The numerical results are based on New England 39-bus system. As shown in Fig.5.15, New England 39-bus system is divided into four areas. The general connection between them is shown in Fig.5.16.

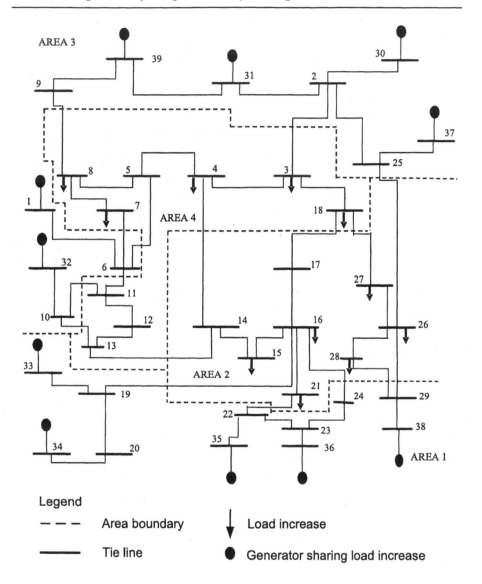

Fig.5.15 New England 39-bus system

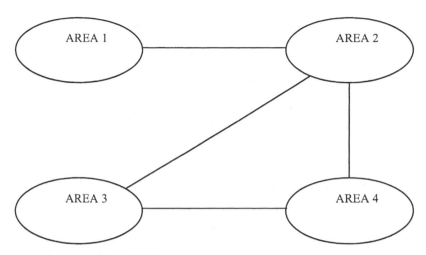

Fig.5.16 Illustration of area connection

There are two ATC scenarios considered in the simulation:

- One Transaction between AREA 1(seller) and AREA 2 (buyer);
- Two simultaneous transactions, one is between AREA 1 (seller) and AREA 2 (buyer), the other is between AREA 3 (seller) and AREA 4(buyer).

The load increase buses and generation sharing buses are listed in Table 5.5.

Table 5.5 Area scenario for New England 39-bus system

AREA #	Load Increase Bus	Generation Sharing Bus
1	None	33, 34, 35, 36, 38
2	15, 16, 21, 26, 27, 28	None
3	None	30, 31, 32, 37, 39
4	3, 4, 7, 8, 18	None

In the above ATC scenarios, the seller's generation should match its associated buyer's load demand. In the case of two simultaneous transactions, the amounts of transaction are proportional to the base case loads of corresponding areas.

The initial operation conditions of the areas are as follows:

Area 1:

Bus	Initial Real Generation (X100MW)	Power Factor
33	6.3	0.95 Leading
34	6.12	0.91 Leading
35	4.88	0.89 Leading
36	6.3	0.95 Leading
38	5.2	0.98 Leading
Total Generation	2880MW	

Area 2:

Bus	Initial Real Load (X100MW)	Power Factor
15	3.2	0.90 Lagging
16	3.294	0.93 Lagging
21	2.74	0.92 Lagging
26	1.39	0.95 Lagging
27	2.81	0.97 Lagging
28	2.06	0.99 Lagging
Total Load	1549 MW	

Area 3:

Bus	Initial Real Generation (X100MW)	Power Factor
30	2.3	0.71 Leading
31	7.23	0.93 Leading
32	6.3	0.92 Leading
37	5.2	0.99 Leading
39	10	0.99 Leading
Total Generation	3103 MW	

Area 4:

Bus	Initial Real Load (X100MW)	Power Factor
3	3.220	0.93 Lagging
4	5.000	0.94 Lagging
7	2.338	0.94 Lagging
8	5.220	0.95 Lagging
18	1.580	0.98 Lagging
Total Load	1736 MW	

Single Transaction:

In the single transaction example below, we determine the coefficients K_{Lk} according to (5.34)

$$K_{Lk}^{(j)} = C \frac{\mu_k^{(j)}}{\sum_{k=1}^{NL_j} \mu_k^{(j)} P_{Lk0}^{(j)}} \frac{\sum_{i=1}^{n} P_{Tij}}{P_{TOTAL}}$$

In (5.34), $\sum_{i=1}^{n} P_{Tij}$ is the total load in Area 2, which has the value 1549 MW; P_{TOTAL} is the total system load of the New England system, which has the value 6141 MW. According to the definition, P_{Lk0} is the existing load at bus k. Thus in this case, we have

$$P_{L15\_0} = 320 \quad P_{L16\_0} = 329.4 \quad P_{L21\_0} = 274$$

$$P_{L26\_0} = 139 \quad P_{L27\_0} = 281 \quad P_{L28\_0} = 206$$

Now the next job is to determine variable C and μ_k. Without loss of generality, we can define scalar $C=1000$. For variable μ_k, it defines load increment scenario in Area 2. And if the load is increased at various buses proportionally to their initial values, we have

$$\mu_{15} = \mu_{16} = \mu_{21} = \mu_{26} = \mu_{27} = \mu_{28} = 1$$

Finally we can calculate all the K_{Lk} in Area 2 with all these parameters:

$$K_{L15} = K_{L16} = K_{L21} = K_{L26} = K_{L27} = K_{L28}$$

$$= 1000 \times \frac{1}{320 + 329.4 + 274 + 139 + 281 + 206} \times \frac{1549}{6141} = 0.1628$$

Similarly, we can also define K_{Gk} according to (5.39). In (5.39), $\sum_{i=1}^{n} P_{Tji}$ is the total generation in Area 1, which has the value 3080 MW; and P_{TOTAL} is the same, 6141 MW. For variable η_k, it defines generation increment scenario in Area 1. And if the generation is increased at various buses proportionally to their initial values, we have

$$\eta_{33} : \eta_{34} : \eta_{35} : \eta_{36} : \eta_{38}$$
$$= P_{G33\_0} : P_{G34\_0} : P_{G35\_0} : P_{G36\_0} : P_{G38\_0}$$
$$= 630 : 612 : 488 : 630 : 520$$
$$= 0.2188 : 0.2125 : 0.1694 : 0.2188 : 0.1806$$

Finally, the coefficients K_{Gk} in area 1 can be calculated as follows:

$$K_{G33} = \frac{0.2188}{1} \times \frac{3080}{6141} = 0.1097$$

$$K_{G34} = \frac{0.2125}{1} \times \frac{3080}{6141} = 0.1066$$

$$K_{G35} = \frac{0.1694}{1} \times \frac{3080}{6141} = 0.0850$$

$$K_{G36} = \frac{0.2188}{1} \times \frac{3080}{6141} = 0.1097$$

$$K_{G38} = \frac{0.1806}{1} \times \frac{3080}{6141} = 0.0906$$

After rescaling K_{Gk} to make the sum equal to 1, we finally get:

$$K_{G33} = 0.2187 \quad K_{G34} = 0.2125 \quad K_{G35} = 0.1695$$
$$K_{G36} = 0.2187 \quad K_{G38} = 0.1806$$

In the single transaction, only area 1 and area 2 are involved. The load and generation conditions at the critical point are shown as follows:

Area 1:

Bus	Generation near the critical point (X100MW)	Power Factor
33	7.766	0.92 Leading
34	6.194	0.92 Leading
35	7.995	0.92 Leading
36	6.853	0.92 Leading
38	10.278	0.96 Leading
Total 1 Generation near the critical point	3909 MW	
Initial total Genera-tion	3080 MW	
Generation Increment	829 MW	

Area 2:

Bus	Critical Load (X100MW)	Power Factor
15	4.879	0.90 Lagging
16	5.023	0.93 Lagging
21	4.178	0.92 Lagging
26	2.119	0.95 Lagging
27	4.285	0.97 Lagging
28	3.141	0.99 Lagging
Total Load at the critical point	2363 MW	
Initial total Load	1549 MW	
Load Increment	814 MW	

Simultaneous Transactions:

By the same way, we can also define the coefficients in simultaneous transactions. In this case, we still use (5.34) and (5.39) to determine these coefficients. And when applying (5.34) and (5.39) to determine Area 1 and Area 2 coefficients, we notice that all the parameters remain the same, thus we have the same coefficients of Area 1 and Area 2 as in the single trans-action case.

As for the Area 3 and Area 4, we only need to know area 4 load and area 3 generation to apply (5.34) and (5.39). In this case, the area 4 load total is 1736 MW, and the area 3 generation total is 3103 MW. Thus, we can calculate the coefficients as follows:

For the load buses:

$$K_{L3} = K_{L4} = K_{L7} = K_{L8} = K_{L18}$$

$$= 1000 \times \frac{1}{322 + 500 + 233.8 + 522 + 158} \times \frac{1736}{6141} = 0.1628$$

For the generator buses:

$$\eta_{30} : \eta_{31} : \eta_{32} : \eta_{37} : \eta_{39}$$

$$= 230 : 723 : 630 : 520 : 1000$$

$$= 0.074 : 0.233 : 0.203 : 0.168 : 0.322$$

$$K_{G30} = \frac{0.074}{1} \times \frac{3103}{6141} = 0.037$$

$$K_{G31} = \frac{0.233}{1} \times \frac{3103}{6141} = 0.118$$

$$K_{G32} = \frac{0.203}{1} \times \frac{3103}{6141} = 0.103$$

$$K_{G37} = \frac{0.168}{1} \times \frac{3103}{6141} = 0.085$$

$$K_{G39} = \frac{0.322}{1} \times \frac{3103}{6141} = 0.163$$

After rescaling K_{Gk} to make the sum equal to 1, we finally get:

$$K_{G30} = 0.073 \quad K_{G31} = 0.233 \quad K_{G32} = 0.204$$

$$K_{G37} = 0.168 \quad K_{G39} = 0.322$$

In the simultaneous transactions, all 4 areas are involved. The load and generation conditions at the critical point are shown as follows:

Area 1:

Bus	Generation near the critical point (X100MW)	Power Factor
33	7.283	0.90 Leading
34	5.809	0.91 Leading
35	7.498	0.87 Leading
36	6.428	0.93 Leading
38	9.64	0.96 Leading
Total Generation near the critical point	3666 MW	
Initial total Generation	3080 MW	
Generation Increment	586 MW	

Area 2:

Bus	Critical Load (100MW)	Power Factor
15	4.371	0.90 Lagging
16	4.499	0.93 Lagging
21	3.742	0.92 Lagging
26	1.899	0.95 Lagging
27	3.838	0.97 Lagging
28	2.814	0.99 Lagging
Total critical Load	2116 MW	
Total Initial Load	1549 MW	
Load Increment	567 MW	

Area 3:

Bus	Generation near the critical point (X100MW)	Power Factor
30	2.785	0.88 Leading
31	8.751	0.89 Leading
32	7.626	0.89 Leading
37	6.287	0.91 Leading
39	12.107	0.96 Leading
Total Generation near the critical point	3756 MW	
Total Generation	3103 MW	
Generation Increment	653 MW	

Area 4:

Bus	Critical Load (X100MW)	Power Factor
3	4.398	0.93 Lagging
4	6.829	0.94 Lagging
7	3.193	0.94 Lagging
8	7.130	0.95 Lagging
18	2.158	0.98 Lagging
Total Critical Load	2371 MW	
Total Initial Load	1736 MW	
Load Increment	635 MW	

After ATC margin is calculated by the given scenarios, ATC margin change with respect to control resources can be calculated by the ATC tracing procedure described in Section 5.3. In the following numerical examples, the ATC margin change with respect to various control actions are given for both single transaction and simultaneous transactions.

Steps involved in tracing ATC as limited by voltage stability is described as follows:

1. Specify a transfer scenario, and calculate scenario coefficients according to (5.34) and (5.39).
2. Equilibrium Tracing Program (EQTP) starts at the current operating point for the initial ATC limit under fixed control configuration and specified transfer scenario.
3. Specify the control scenario that describes the change of control configuration or contingencies.
4. Change control parameter to new value βi, find the new initial starting point as shown in Fig. 5.1.
5. Use EQTP to trace the new ATC limit λ_i starting from the initial starting point in step 4.
6. Go to step 4 unless some control variables hit limits.

In the following sections, several numerical examples are given to show the ATC tracing procedure with respect to various controls. Two transaction scenarios; single and simultaneous transactions are considered. In the single transaction, power is transferred from area 1 to area 2, and the initial ATC limit is 814MW; while in the simultaneous transactions, power is transferred both from area 1 to area 2 and from area 3 to area 4, and the initial ATC limits are 567MW and 635MW respectively. Once the initial

ATC limit is calculated, control actions such as load relief, reactive power support, V_{ref} change are applied to directly trace the new ATC limit. The variations of ATC with respect to control actions are shown in the various figures below.

5.8.2 Emergency transmission load relief

In certain extreme conditions, transmission load relief (TLR) procedure is implemented to relieve overloading in the transmission system. Simulation has been done to find out the effectiveness of implementing TLR on certain buses.

5.8.2.1 Single transaction case

Fig.5.17 demonstrates the SNB related ATC margin change with TLR implemented at bus 7. At the base case, The SNB related ATC margin between AREA 1 and AREA 2 is 814MW. The SNB related ATC margin between AREA 1 and AREA 2 reaches its maximum 864MW when 140MW load is shed.

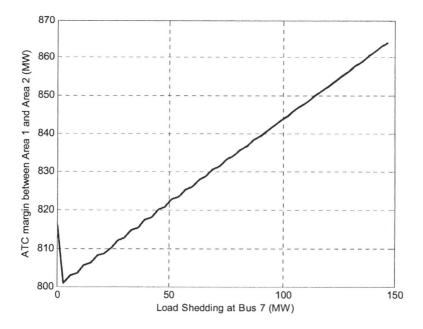

Fig.5.17 ATC margin vs. TLR implemented at bus 7 (single transaction)

5.8.2.2 Simultaneous transaction case

For the simultaneous transaction case, the amounts of power transfer for two transactions are proportional to each other according to the scenario setting. The seller's generation increase in AREA 1/AREA 3 matches the buyer's load increase in AREA 2/AREA 4 respectively. Fig.5.18 shows the voltage stability (SNB) related ATC margin change for both transactions. Along with the load shedding at bus 27, the SNB related ATC margin for both transactions increase from 567MW/635MW to 670MW/738MW respectively. The margin tracing curve is smooth and close to linear.

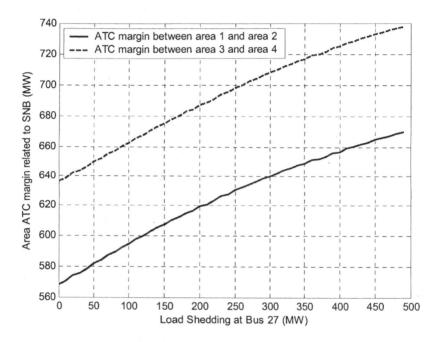

Fig.5.18 SNB related ATC margin vs.TLR implemented at bus 27 (simultaneous transaction case)

5.8.3 Reactive power Support

5.8.3.1 Single transaction case

Fig.5.19 shows the SNB related ATC margin change between AREA 1 and AREA 2 as shunt capacitance increases at bus 7. The ATC margin tracing curves show the highly nonlinear characteristics and some "jumps" because of generators hitting their limits.

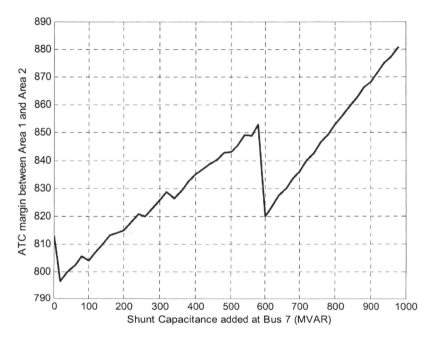

Fig.5.19 ATC margin vs. shunt capacitance at bus 7 (single transaction case)

5.8.3.2 Simultaneous transaction case

Fig.5.20 demonstrates the SNB related ATC margin change for simultaneous transactions as the shunt capacitance increases at bus 21. In Fig.5.20, the drop in stability margin at 300 MVAr shunt capacitance level is caused by generator 30 hitting its I_a and V_r limits.

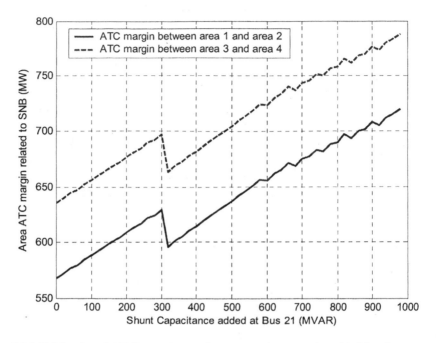

Fig. 5.20 SNB related ATC margin vs. shunt capacitance at bus 21 (simultaneous transaction case)

5.8.4 Control combination

The control scenario could be any combination of control parameters. The direct ATC tracing method can trace margin boundary with respect to the multi-control parameter space.

Fig.5.21 shows how the ATC margin changes with respect to a control scenario: At each step V_{ref} of generator 39 increases by 0.001 p.u; shunt capacitance at bus 31 increases by 0.1 p.u. and load shedding at bus 4 by 0.1 p.u. The control scenario simulates the total effect of secondary voltage regulation, as well as reactive power support and emergency TLR scheme.

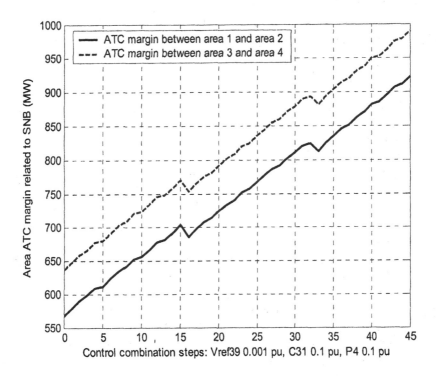

Fig.5.21 SNB related ATC margin vs. control combination (simultaneous transaction case)

5.9 Conclusion

In this chapter, the general framework of stability margin boundary tracing described in section 5.3 is reformulated to trace the voltage/oscillatory stability related ATC margin. The SNB related ATC margin boundary can be identified and traced along any control scenario combined with any given load/generation increase scenario.

The aim here is to demonstrate the application of continuation based technique for transfer capability calculation. Literature related to transfer margin and ATC can be found in references [10-12]. Ref [13] provides an excellent tutorial introduction to ATC with examples.

References

[1]   Van Cutsem, T., An approach to corrective control of voltage instability using simulation and sensitivity. IEEE Trans. on power systems, Vol. 10: 616-622, 1995

[2]   Flueck, A.J., Dondeti, J.R., "A new continuation power flow tool for investigating the nonlinear effects of transmission branch parameter variations,"*IEEE Trans. on power systems*, vol.15, no.1, pp.223-227, Feb. 2000

[3]   Zhou, Y., *Manifold based voltage stability boundary tracing, margin control optimization and time domain simulation*, Ph.D. thesis, Iowa State University, Ames, IA, 2001

[4]   Dai, R., Rheinboldt, W.C., "On the computation of manifolds of foldpoints for parameter-dependent problems," *SIAM J. Numer. Anal.*, Vol. 27, no.2, pp. 437-446, 1990

[5]   Feng, Z., Ajjarapu, V., Maratukulam, D. J. , Identification of voltage collapse through direct equilibrium tracing. IEEE Trans. on Power System, Vol. 15, pp. 342-349, 2000

[6]   Van Cutsem, T., Vournas, C., *Voltage stability of Electric Power Systems*, Kluwer Academic Publishers, Norwell, MA, 2003

[7]   http://www.nerc.com/~filez/reports.html. Transmission transfer capability task force, Available transfer capability definitions and determination. North American electric reliability council, Princeton, New Jersey, 1996

[8]   Austria, R. R., Chao, X. Y., Reppen, N. D., Welsh, D. E., Integrated approach to transfer limit calculations. IEEE computer applications in power: 48-52, 1995

[9]   Wang, G., Voltage Stability Based ATC and Congestion Management for Simultaneous Multi-Area Transactions, M.S. Thesis, July 1998

[10]   North American Electric Reliability Council, "Transmission Transfer Capability", May 1995.

[11]   Ilic, M.D., Yoon, Y.T., Zobian, A., "Available transmission capacity (ATC) and its value under open access," IEEE Trans. on Power System, Vol. 12, no.2, pp. 636-645, 1997

[12]   Greene, S., Dobson, I., Alvarado, F.L., "Sensitivity of transfer capability margins with a fast formula," IEEE Trans. on Power System, Vol. 17, no.1, pp. 34-40, Feb.2002

[13]   http://www.pserc.cornell.edu/tcc/

6 Time Domain Simulation

6.1 Introduction

In the previous chapters the continuation based approaches to study steady state aspects of voltage stability are presented. We still need time domain simulations to capture the transient response and timing of control actions. The time domain response can capture the evolvement of the instability process to provide the timing issues of controls. To capture the transient response a set of differential and algebraic equations (DAE) are numerically solved. Power systems networks typically include a large number of dynamic and static components, where each individual component may need several differential and algebraic equations to represent, thus the total number of differential and algebraic equations of a real power system can be quite large.

The time constants of these power system components vary in a large range and leads to stiffness in the system. The numerical methods for time domain simulations may produce wrong results for stiff systems due to error accumulation in the step-by-step numerical integration. A considerable amount of progress has been made in the power system literature [1-10] to solve large scale stiff power systems. This chapter describes a promising decoupled simulation method published in [22] to address the stiff problems in power systems. Sections 6.2 to 6.4 are based on this paper and some of the material in the paper is reproduced for continuity. Finally this chapter tries to provide information that is related to computational aspects of the short term and long-term time scales.

6.2 Explicit and Implicit Methods

Time domain simulations are needed for numerical integration methods. These integration methods can be classified into two categories: explicit methods and implicit methods. The explicit methods involve fixed point iteration and can lead to faster solution. However the explicit methods have numerical stability problem when dealing with stiff problems. In contrast the implicit methods are stable but slow.

Consider a general ODE system with a given initial condition as described by (6.1):

$$\begin{cases} \dot{x} = f(x) \\ x(0) = x_0 \end{cases} \tag{6.1}$$

There is a well established mathematical literature to solve the above initial value problem (IVP) [11-15]. The following sections provide information related to the most commonly used explicit and the implicit methods.

6.2.1 Explicit method

Explicit methods typically replace the ordinary differential equations by nonlinear recursive mappings

$$x_{k+1} = g(x_k) \tag{6.2}$$

For a given initial value x_0, the states $x_1, x_2, \cdots$ are generated as long as they remain in the domain of definition of the mapping.

Some of the methods under this category include: forward Euler, explicit Runge-Kutta methods, and Adams-Bashforth methods.

The forward Euler method is formulated as:

$$x_{k+1} = x_k + hf(t_k, x_k) \tag{6.3}$$

The 4th-order Runge-Kutta method is formulated as:

$$K_1 = f(t_k, x_k)$$
$$K_2 = f(t_k + h/2, x_k + hK_1/2)$$
$$K_3 = f(t_k + h/2, x_k + hK_2/2) \qquad\qquad (6.4)$$
$$K_4 = f(t_k + h, x_k + hK_3)$$
$$x_{k+1} = x_k + h(K_1 + 2K_2 + 2K_3 + K_4)/6$$

As you can see from (6.3) and (6.4), the current solution is generated from previous results. Thus explicit methods are very efficient with fixed point iteration techniques.

6.2.2 Implicit method

Implicit methods need both the current state and past state to solve the initial value problem.

Backward Euler method, Trapezoidal method, implicit Runge-Kutta method, Adams-Moulton methods, Backward Differential Formulae methods are part of the whole family of the implicit methods.

The backward Euler method is formulated as:
$$x_{k+1} = x_k + hf(t_{k+1}, x_{k+1}) \qquad\qquad (6.5)$$
The Trapezoidal method is formulated as:
$$x_{k+1} = x_k + h[f(t_k, x_k) + f(t_{k+1}, x_{k+1})]/2 \qquad\qquad (6.6)$$

As you can see from (6.5) and (6.6), implicit methods involve solving a set of nonlinear equations. Here the next state cannot be obtained directly, and Newton method is usually used to solve the nonlinear equations. However, the implicit methods have better numerical stability properties than explicit methods despite their slow computational performance.

6.2.3 Stiffness and Numerical Stability

In the introduction we mentioned the term stiffness. The stiffness in a system can be due to the components with vastly different time scales. From the mathematical viewpoint the stiffness is associated with the existence of both large and small eigenvalues. The quotient of the largest and the smallest eigenvalues can be considered as the stiffness ratio to measure the degree of stiffness [14].

Stiffness creates problems during the integration. This not only leads to convergence problems but also can produce the wrong outcome. We want the numerical method to correctly identify whether a particular system subject to some form of disturbance is stable or unstable. There is a possibility that the numerical solution may indicate unstable behavior for the case where the actual system is stable and vice versa. We need some form of confidence in the quality of the numerical result. In the mathematical literature [15] [16], the step size needed to guarantee numerical stability is stated.   In the case of explicit methods we may need a significant reduction of the step size to maintain numerical stability such that the step size is smaller than the step size needed to represent the solution accurately. The required step size for explicit methods to guarantee numerical stability may be too small for practical implementation [15] [16]. However the step size control to maintain numerical stability for stiff systems can be achieved through the implicit methods.

In [22] a simple example given in [14] is used to show the difference between explicit and implicit methods for stiff systems. The example is repeated here.   It is a simple two variable example and can demonstrate the basic concepts.

$$\begin{cases} \dot{x}_1 = -100x_1 + x_2 \\ \dot{x}_2 = -0.1x_2 \end{cases}$$

(6.7)

    We have chosen forward Euler and Trapezoidal methods as explicit and implicit methods respectively.

The initial value is $(-3, -1)^T$, and the step size is chosen as 0.1. The eigenvalues of (6.7) by inspection are -100 and -0.1(as you can see there is a large difference between the eigenvalues indicating stiff system). Since the real parts are negative, the system trajectory converges to the origin as time goes to infinity. The results by forward Euler method and Trapezoidal methods are shown in Fig. 6.1a, b respectively. Forward Euler method (Fig. 6.1a) is providing wrong information, in contrast with the asymptotic behavior of the true solution. To capture the corrector behavior with Forward Euler, the step size should be reduced to below 0.02. Trapezoidal method (Fig. 6.1b) is providing proper stable behavior.

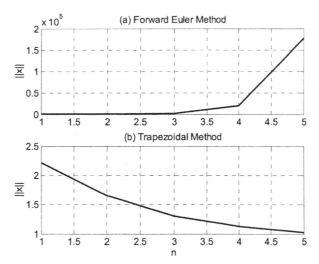

Fig. 6.1. The simulation results by explicit and implicit methods [22]

In the mathematical literature [11-15], to study the stability of numerical methods the concepts of A-stability and stability domain are proposed.
Stability Domain: Suppose that a given numerical method is applied with a step size $h>0$ to the linear test system $x' = \lambda x$, the stability domain of the underlying numerical method is the set of all numbers $h\lambda$ such that x_n approach zero as $n \to \infty$. In other words, the stability domain is the set of all $h\lambda$ for which the correct asymptotic behavior is recovered, provided that the linear system is stable.

A -stability: A method is *A-stable* if x_n approach zero as $n \to \infty$ for all values of the step size h when this method is applied to the equation $x' = \lambda x$ for all $\lambda \in C$ with $Re(\lambda)<0$. Note that for this equation, the exact solution also goes to zero. In other words for $Re(\lambda)<0$ the solution of corresponding differential equations should be stable for any positive value of h. It implies that the stability domain includes the whole left half plane. Whether a method is *A-stable* or not can be judged from the stability domain. The stability domains of the forward Euler and Trapezoidal method are shown in Fig.6.2. The forward Euler method is not *A-stable* while Trapezoidal method is *A-stable*. It is proven that no explicit Runge-Kutta method may be *A-stable* [14]. In general, only implicit methods may be *A-stable*.

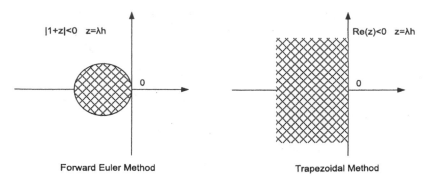

Fig.6.2 Stability Domain of Forward Euler and Trapezoidal Methods

The drawback of A-stability is that the stability domain may include part of the right half plane, thus the real unstable phenomena will be simulated as a stable one. The spurious damping is called hyper stability [1]. Hyper stability can be avoided by reducing the step size during the simulation on the basis of the experience of the end user or the evaluation of eigenvalues.

The following section describes a time domain algorithm [22] which takes advantage of the desirable properties of implicit and explicit methods.

6.3 Decoupled Time Domain Simulation [22]

Since power systems are stiff problems, implicit methods are commonly used to simulate the dynamic behavior. Each integration step of a stiff equation involves the solution of a nonlinear equation which leads to a set of linear problems involving the Jacobian of the system. As a result, the methods for solving stiff systems spend most of the time solving systems of linear equations. The decoupled time domain simulation by Yang, in [22], aims to reduce computational burden of the traditionally implicit methods. In fact, the numerical stability properties of the time domain simulation algorithms are determined by the eigenvalues of the linearized matrix, and frequently the eigenvalues which cause stiff problems are only a small portion of the whole spectra. It seems inefficient to solve these problems only with implicit methods. If the problem can be partitioned into a stiff part and a non stiff part such as

$$\begin{cases} \dot{x}_s = f_s(x_s, x_n) \\ \dot{x}_n = f_n(x_s, x_n) \end{cases} \tag{6.8}$$

where x_s, x_n are stiff and non stiff variables, and f_s, f_n are stiff and non stiff equations, the system can be treated with an implicit method for the stiff components and an explicit method for the non stiff parts [12].

For the numerical stability, it is required that eigenvalues are located inside the stability domain to yield convergence behavior. If some eigenvalues are outside the stability domain of explicit methods, numerical stability may not be revealed by dynamic simulation. However, the numerical results can be corrected by treating those outside eigenvalues differently. The decoupled method is based on the idea of separating stiff eigenvalues from the others.

From the geometric viewpoint, the solutions of the ODE and DAE systems are points or vectors in the multi-dimension space. This space can be divided into two or more subspaces and the solution vectors can be decomposed into corresponding two or more sub-vectors in each subspaces. Thus, by decomposing the space into a number of small subspaces, the solution vectors can be divided into sub-vectors and the original ODE and DAE systems can be decoupled into several small dimension systems.

Let P be the invariant subspace corresponding to the m eigenvalues which are outside the stability domain of an explicit method and let Z_1 be an orthonormal basis in P. Thus Z_1 is n × m matrix which satisfies the following conditions:

$$AZ_1 = Z_1 \Lambda_1 \tag{6.9}$$

$$Z_1^T Z_1 = I_m \tag{6.10}$$

where I_m is m × m identity matrix, and Λ_1 is a square matrix with m eigenvalues which are outside the stability domain.

Furthermore, there exists an orthogonal complement Q such that $Q = P^{\perp}$. And let Z_2 be the orthonormal basis in Q, then

$$Z_2^T Z_2 = I_{n-m} \tag{6.11}$$

And since Q is an orthogonal complement, it follows that

$$Z_1^T Z_2 = 0 \quad Z_2^T Z_1 = 0 \tag{6.12}$$

Therefore, the space can be represented by the direct sum of P and Q where Z_1 and Z_2 are the corresponding basis respectively. Moreover, m×m dimension matrix $Z_1 Z_1^T$ and $(n-m)\times(n-m)$ dimension matrix $Z_2 Z_2^T$ are the orthogonal projectors into the two subspaces according to the definition in [21].

Because $Z_1 Z_1^T$ and $Z_2 Z_2^T$ are the orthogonal projectors, any vector in the full space can be projected into two subspaces by multiplying the projectors on the left. In other words, once the projections in these two subspaces are known, the original vector in the full space can be recovered. For each vector x in R^n space, there exists a vector $p \in R^m$ and $q \in R^{n-m}$ such that $x = Z_1 p + Z_2 q$, and it can be proved by setting $p = Z_1^T x$ and $q = Z_2^T x$.

Since the vector in the original n-dimension space can be decomposed into the sum of two small dimension vectors, the original system can be split into two sub systems according to [17] [18]:

$$f^P(p,q) = Z_1^T f(Z_1 p + Z_2 q) \tag{6.13}$$

$$f^Q(p,q) = Z_2^T f(Z_1 p + Z_2 q) \tag{6.14}$$

And the ODE system equations can be decoupled into two systems

$$\begin{cases} \dot{p} = f^P(p,q) = Z_1^T f(Z_1 p + Z_2 q) \\ \dot{q} = f^Q(p,q) = Z_2^T f(Z_1 p + Z_2 q) \end{cases} \tag{6.15}$$

By solving the above decoupled equations, p and q can be calculated separately, and the original states are given as $x = Z_1 p + Z_2 q$.

For the decoupled systems, the second set of equations $Z_2^T f(Z_1 p + Z_2 q)$ has the derivative $Z_2^T f_x Z_2$. It is desirable if the eigenvalues are still in the stability region of the explicit methods.

Proposition 1: The matrix $Z_2 f_x Z_2$ has the remaining n-m eigenvalues $\lambda_{m+1}, \cdots, \lambda_n$.
Proof is given in [17] [18].   □

Equation (6.15) has the desired form as (6.8) and the all the eigenvalues of the second equation set are inside the stability domain. Therefore, an explicit method can be applied to solve the second set of equations and an implicit method can be applied to solve the first set of equations. To eliminate the need for Z_2, let $v = Z_2 q$, thus

$$\dot{v} = Z_2 Z_2^T f(Z_1 p + v) = (I - Z_1 Z_1^T) f(Z_1 p + v)$$

The new system is

$$\begin{cases} \dot{p} = Z_1^T f(Z_1 p + v) \\ \dot{v} = (I - Z_1 Z_1^T) f(Z_1 p + v) \end{cases} \tag{6.16}$$

Decoupled Method for DAE

The simulation of Differential Algebraic Equation systems involves solving of a set of differential equations and a set of algebraic equations simultaneously. The solutions of differential equations and algebraic equations can be obtained either separately or simultaneously [19]. The decoupled method can be applied to DAE systems in the similar way to ODE systems. To demonstrate the approach, forward Euler method is chosen as an example of explicit method and Trapezoidal method as an example of the implicit method. Also we denote the number of differential equations as n, the number of the algebraic equations as l, and the dimension of stiff invariant subspace P as m.

Decoupled Forward Euler-Trapezoidal Method for DAE

Similar to the ODE system, the DAE system

$$\begin{cases} \dot{X} = F(X,Y) \\ 0 = G(X,Y) \end{cases}$$

can be decomposed into the following form

$$\begin{cases} \dot{p} = Z_1^T F(Z_1 p + v) \\ \dot{v} = (I - Z_1 Z_1^T) F(Z_1 p + v) \\ 0 = G(p, v, Y) \end{cases} \tag{6.17}$$

The initial conditions are given as

$$\begin{cases} X(0) = X_0 \\ Y(0) = Y_0 \\ p(0) = Z_1^T X_0 \\ v(0) = (I - Z_1 Z_1^T) X_0 \end{cases} \tag{6.18}$$

The decoupled forward Euler-Trapezoidal method is formulated as:

$$v_{k+1} = v_k + h(I - Z_1 Z_1^T) F(Z_1 p_k + v_k)$$

$$p_{k+1} = p_k + h Z_1 [(F(Z_1 p_k + v_k) + F(Z_1 p_{k+1} + v_{k+1}))]/2 \tag{6.19}$$

$$0 = G(p_{k+1}, v_{k+1}, Y_{k+1})$$

The first set of equations can be solved via fixed point iteration, and the second and third sets of the equations actually are nonlinear and Newton method is needed to solve them. The second and third equation sets are reformulated as:

$$\begin{cases} p_{k+1} - \dfrac{1}{2} h Z_1 F(Z_1 p_{k+1} + v_{k+1}) = p_k + \dfrac{1}{2} h Z_1 F(Z_1 p_k + v_k) \\ g(p_{k+1}, v_{k+1}, Y_{k+1}) = 0 \end{cases} \tag{6.20}$$

where the unknowns are p_{k+1} and y_{k+1}.

Compared with above, the full implicit method needs to solve the following nonlinear equation set:

$$\begin{cases} x_{k+1} - h f(x_{k+1})/2 = x_k + h f(x_k)/2 \\ g(x_{k+1}, y_{k+1}) = 0 \end{cases} \tag{6.21}$$

The dimension of the full implicit method is n+1 while the dimension of the decoupled system is m+1. Since m<<n, the dimension of the nonlinear systems can be significantly reduced.

Identification of Stiff Invariant Subspace Basis Z_1

To identify the basis Z1 of invariant subspace P, it is necessary to identify the eigenvectors of corresponding eigenvalues. As for the forward Euler method, the stability domain is a circle which has the center (-1/h, 0) and radius 1/h. Thus the stiff invariant subspace is associated with eigenvalues outside the circle. It is difficult to directly identify the invariant subspace if the center of the circle is not the origin. The remedy is to shift the original eigenvalues to the right direction so that the origin becomes the center of the circle. One can show that if $\lambda_1, \cdots, \lambda_n$ are eigenvalues of n-by-n matrix A, then $\lambda_1 + 1/h, \cdots, \lambda_n + 1/h$ are the eigenvalues of matrix $A + I/h$ (Let λ_i, x_i be the corresponding eigen-pair to matrix A, thus $Ax_i = \lambda_i x_i$. Then $Ax_i + x_i/h = \lambda_i x_i + x_i/h$, that is $(A + I/h)x_i = (\lambda_i + 1/h)x_i$. Therefore $\lambda_i + 1/h$ is the eigenvalue of $A + I/h$).

Now the original problem is converted into the new problem to find out the eigenvalues outside a circle of the matrix $A + I/h$. These eigenvalues can be computed efficiently by the Arnoldi method.

Major steps involved in the decoupled simulation:

1. Calculate equilibrium
2. Calculate linearized matrix A
3. Obtain linear transformed matrix
4. Identify transformed matrix stiff eigenvalues and eigenvectors with largest moduli by dominant eigenvalue algorithm
5. Update non-stiff subsystem by forward Euler method
6. Update stiff subsystem by Trapezoidal method
7. Continue steps 5 and 6 until a specified simulation time is reached.

6.4 Numerical Examples

First a simple two bus example introduced in the previous chapters is used to demonstrate various steps involved in implementing the method.

6.4.1 Two bus system

Consider the two bus example introduced in the previous chapters at page 54. With the two axis model, there are in total 9 states. The differential and algebraic equations describing the system are given below.

F part:

$$\dot{\delta_1} = (\omega_1 - \omega_m)\omega_0$$

$$\dot{\omega_1} = M_1^{-1}[P_{m1} - D_1(\omega_1 - \omega_m) - (E_{q1}' - X_{d1}'I_{d1})I_{q1}$$
$$- (E_{d1}' + X_{q1}'I_{q1})I_{d1}]$$

$$\dot{E_{q1}'} = T_{d01}^{-1}[E_{fd1} - E_{q1}' - (X_{d1} - X_{d1}')I_{d1}]$$

$$\dot{E_{d1}'} = T_{q01}^{-1}[-E_{d1}' + (X_{q1} - X_{q1}')I_{q1}]$$

$$I_{d1} = [R_{s1}E_{d1}' + E_{q1}'X_{q1}' - R_{s1}V_1\sin(\delta_1 - \theta_1) - X_{q1}'V_1\cos(\delta_1$$

$$I_{q1} = [R_{s1}E_{q1}' - E_{d1}'X_{d1}' - R_{s1}V_1\cos(\delta_1 - \theta_1) - X_{d1}'V_1\sin(\delta_1$$

$$A_1 = R_{s1}^2 + X_{d1}'X_{q1}'$$

$$\dot{E_{fd1}} = T_{e1}^{-1}[V_{r1} - [S_{e1}(E_{fd1})]E_{fd1}]$$

$$\dot{V_{r1}} = T_{a1}^{-1}[-V_{r1} + K_{a1}(V_{ref1} - V_1 - R_{f1})]$$

If $V_{r1,min} \le V_{r1} \le V_{r1,max}$, $V_{pss1} = 0$ (at steady state),

$$\dot{R_{f1}} = T_{f1}^{-1}[-R_{f1} - [K_{e1} + S_{e1}(E_{fd1})]K_{f1}E_{fd1}/T_{e1} + K_{f1}V_{r1}/T_{e1}$$

$$\dot{P_{m1}} = T_{ch1}^{-1}(\mu_1 - P_{m1})$$

$$\dot{\mu_1} = T_{g1}^{-1}[P_{gs1} - (\omega_1 - \omega_{ref})/R_1 - \mu_1] \qquad \text{if } \mu_{1,min} \le \mu_1 \le \mu_{1,max}$$

$$P_{gs1} = P_{gs1}^0 (1 + K_{g1}\lambda)$$

G part:

$$0 = P_{g1} - P_{t1}$$
$$0 = Q_{g1} - Q_{t1}$$
$$0 = 0 - P_{l20}(1 + \lambda K_{lp2}) - P_{t2}$$
$$0 = 0 - Q_{l20}(1 + \lambda K_{lq2}) - Q_{t2}$$

where,

$$\begin{cases} P_{g1} = I_{d1}V_1 \sin(\delta_1 - \theta_1) + I_{q1}V_1 \cos(\delta_1 - \theta_1) \\ Q_{g1} = I_{d1}V_1 \cos(\delta_1 - \theta_1) - I_{q1}V_1 \sin(\delta_1 - \theta_1) \end{cases}$$

$$\begin{cases} P_{t2} = Y_{21}V_1V_2 \cos(\theta_2 - \theta_1 - \varphi_{21}) + Y_{22}V_2^2 \cos(\varphi_{22}) \\ Q_{t2} = -Y_{21}V_1V_2 \sin(\theta_2 - \theta_1 - \varphi_{21}) - Y_{22}V_2^2 \sin(\varphi_{22}) \end{cases}$$

$$|Y_{21}| = 10, \varphi_{21} = 90°, \varphi_{22} = -90°$$

Parameters used in this example are given in Appendix A.

Given the initial operating condition:

$$P_{l20} = 1.4 \, p.u., Q_{l20} = 0, P_{Gs1}^0 = 1.4 \, p.u., Q_{Gs1}^0 = 0.32, K_{g1} = 1.0, K_{lp2} = 1.0$$

In the time domain simulation, we simulate the system behavior under load increment. The load at bus 2 is increased by 0.5 percent per second during the initial 10 seconds, and then stops increasing for the duration of the simulation (20 seconds). Before simulation, we first calculate the invariant subspace Z_1 which is the orthonormal transform of eigenvectors. In this case, the eigenvalues associated with stiff subspace are those outside the circle in the λ plane. These eigenvalues are identified as: -0.1097 + 3.1654i, -0.1097 - 3.1654i. The eigenvectors of the pair of complex eigenvalues are shown as:

$$\begin{array}{cc}
0 & 0 \\
0.0000 + 0.0000i & 0.0000 - 0.0000i \\
-0.0183 + 0.0007i & -0.0183 - 0.0007i \\
-0.0019 + 0.0053i & -0.0019 - 0.0053i \\
0.1206 - 0.3302i & 0.1206 + 0.3302i \\
0.9359 & 0.9359 \\
0.0062 - 0.0080i & 0.0062 + 0.0080i \\
-0.0000 + 0.0000i & -0.0000 - 0.0000i \\
-0.0000 - 0.0000i & -0.0000 + 0.0000i
\end{array}$$

The orthonormal form of these two eigenvectors will give the basis of invariant subspace Z_1, which is:

$$\begin{array}{cc}
0 & 0.0000 \\
-0.0000 & -0.0000 \\
0.0193 & 0.0007 \\
0.0023 & -0.0159 \\
-0.1452 & 0.9890 \\
-0.9892 & -0.1454 \\
-0.0070 & 0.0235 \\
0.0000 & -0.0000 \\
0.0000 & 0.0000
\end{array}$$

Once Z_1 is known, we can solve the following equations to get the system behavior response subject to load increment.

$$\begin{cases}
v_{k+1} = v_k + h(I - Z_1 Z_1^T) F(Z_1 p_k + v_k) \\
p_{k+1} = p_k + h Z_1^T [F(Z_1 p_k + v_k) + F(Z_1 p_{k+1} + v_{k+1})]/2 \\
0 = g(p_{k+1}, v_{k+1}, Y_{k+1})
\end{cases}$$

The steps involved are the following:
1. Update the variable v_{k+1} directly from the last step value

$$v_{k+1} = v_k + h(I - Z_1 Z_1^T) F(Z_1 p_k + v_k)$$

2. Once the variable v_{k+1} is calculated from step 1, we can get the solution for the other two variables p_{k+1} and Y_{k+1} by solving the following equations.

$$p_{k+1} = p_k + hZ_1^T[F(Z_1p_k + v_k) + F(Z_1p_{k+1} + v_{k+1})]/2$$
$$0 = g(p_{k+1}, v_{k+1}, Y_{k+1})$$

Since it is a nonlinear equation set, Newton method can be used to find the solution for p_{k+1} and Y_{k+1}.

Now we show some intermediate steps of the time domain simulation results. Suppose we already got the solution at k = 19 (at the time of 0.475 second), and we want to find the solution at k+1= 20 (at the time of 0.5 second).

At step k=19, the initial values are:

Y(k) =
[0.9992   -0.0012   0.9891   -0.1436]

X(k) =
[0.8620   1.0000   0.9276   0.4056   2.6477   2.6617
   0.0001   1.4005   1.4032]

v(k) =
[0.8620   1.0000   0.9839   0.4479   0.0110   0.0191
   -0.0732   1.4005   1.4032]

p(k) =
[-2.9986   2.2258]

To get the next step(k+1=20) solution, we first solve the solution of $v(k+1)$ from the previous step result $v(k)$ and $p(k)$.

The solution of $v(k+1)$ is:

v(k+1)=
[0.8620   1.0000   0.9840   0.4479   0.0110   0.0191
   -0.0732   1.4005   1.4034]

After v(k+1) is solved, we can substitute it into the second and third equation sets to get p(k+1) and Y(k+1). Here v(k), v(k+1), p(k) are known variables, and p(k+1) and v(k+1) are unknowns. The results are shown below:

The solution of p(k+1) is:

p(k+1) =

  [-2.9996   2.2261]

The solution of Y(k+1) is:

  [0.9992   -0.0013   0.9891   -0.1437   ]

X(k+1) = Z1*p(k+1) + v(k+1)

[0.8620   1.0000   0.9276   0.4056   2.6482   2.6627
   0.0001   1.4005   1.4034]

The voltage of bus 2 during the simulation is plotted in Fig.6.3.

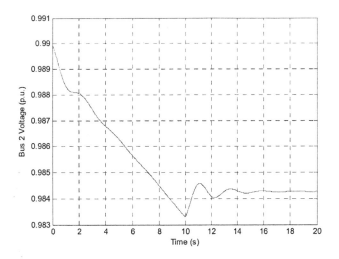

Fig.6.3 Voltage at bus 2 with respect to time (p.u.)

6.4.2 New England 39-bus system

The decoupled forward Euler-Trapezoidal method, full explicit forward Euler method, and full implicit Trapezoidal method are applied to New England system which has 39 buses and 10 generators [22]. The generators are represented by two-axis model as in [19], and exciter and governor model are the same as in [20]. There are 9 states for each generator, and the total number of differential states and algebraic states are 90 and 78 respectively. The step size during the simulation is chosen as 0.025 second. The stiff invariant subspace is calculated at the initial state with dimension as 19, thus the dimension of the nonlinear equation system is 97 for the decoupled method, while the dimensions of the nonlinear systems are 78 and 168 for the explicit method and implicit method respectively.

Time Domain Simulation with Line Trip Contingency

The contingency is transmission line trip between bus 6 and bus 7 at 0.05 second, and the simulation duration is 20 seconds. The actual post-disturbance behavior is that the system stability can be maintained. The results of decoupled method and full implicit method yield stable cases; however, the full explicit method fails to give the correct answer.

Full explicit method (pure forward Euler method) diverges at about 1.1 second as shown in Fig.6.4a. Before explicit method diverges, an oscillatory behavior can be observed t which is only due to numerical error instead of real system response.

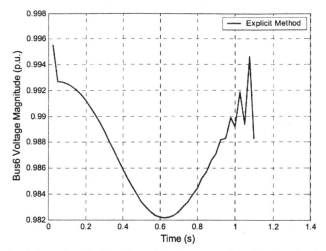

Fig.6.4a. Simulation Result by Forward Euler Method [22]

The simulation results of decoupled method and full implicit method give the stable system behaviors as shown in Fig.6.4b. Both methods give stable post disturbance behavior and the results from two methods match very well.

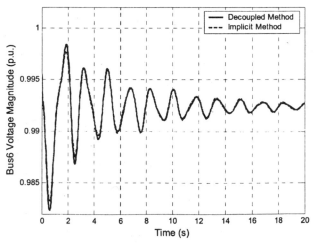

Fig.6.4b. Simulation Results by Decoupled and Full Implicit Methods [22]

The computational time of decoupled method, explicit method and full implicit method is shown in Table 1 (N/A means dynamic simulation cannot be finished due to numerical divergence). It shows that decoupled method requires much less time than implicit method to finish the dynamic simulation.

Table 1 Computational Time for Line Trip [22]

Methods	CPU Time (s)
Explicit Method	N/A
Implicit Method	745
Decoupled Method	405

Long-Term Time Domain Simulation for gradual increase of load

System loads are increased from 6141MW by 0.5 percent per second to capture the long term system instability behavior. The simulation result is shown in Fig. 6.5.

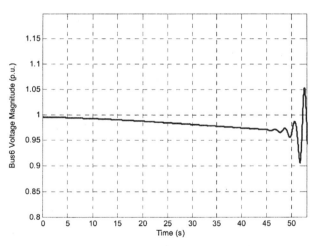

Fig.6.5. Long-Term Simulation Result

Reference [22] provides additional cases with 39 bus system as well IEEE 118 bus system

The decoupled method [22] presented in this chapter can capture the short term as well as long term aspects of voltage instability. In this approach stiff and non stiff subspaces are separated with each other and are treated with different methods. As a result, the decoupled method combines the advantages of pure explicit and pure implicit methods to achieve both numerical stability and efficiency. However as with any time domain methods, it still takes a considerable amount of CPU time to capture the relevant system behavior, especially for long term simulation.   To capture the long term voltage stability behavior other effective methods are presented in the literature to reduce the computational burden. The most widely used approach is Quasi Steady State (QSS) Simulation [23-26]. In the next section the QSS approach in the frame work of continuation based equilibrium tracing is presented.

6.5 Quasi-Steady-State Simulation (QSS)

Quasi-Steady-State (QSS) is the approximation to the system behavior under certain conditions. It assumes that: the fast subsystem is infinitely fast and can be replaced by its equilibrium equation when dealing with the slow subsystem. Conversely, the fast dynamics can be approximated by considering the slow variables as practically constant during the fast transients. This leads to a significantly simpler analysis of both subsystems [23-27].

To demonstrate the methodology, a scenario is first defined as follows:

The post-contingency long-term load characteristic intersects the system PV curve as shown in Fig. 6.6.

In this scenario (Fig.6.6), the short-term load characteristic is fully restored to the long-term load characteristic. At first, the system is at its pre-contingency operating point A. Due to the short-term load characteristic, the system jumps to A' just after the contingency. Each point on the post-contingency PV curve is the short-term equilibrium before the complete restoration (B) and the long-term equilibrium afterwards. Once the restoration is fully achieved, the load is increased to dominate the system's evolution. In Fig. 6.6, the long-term saddle node bifurcation point (SNB) C needs to be identified during the equilibrium tracing in order to obtain the information of how much active power margin the system has from point B.

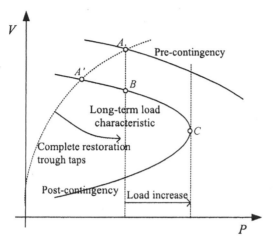

Fig. 6.6.   Demonstration for defined Scenario

6.5.1 Problem Formulation

A general modeling relevant to voltage stability in different time scales is described from the following set of equations [23]: This includes differential, algebraic and discrete sets of equations.

$$\dot{X} = F(X,Y,z_D,z_C,\lambda) \tag{6.22}$$

$$0 = G(X,Y,z_D,z_C,\lambda) \tag{6.23}$$

$$z_D(k^+) = h_D(x(k^-),y(k^-),z_D(k^-),\lambda(k)) \tag{6.24}$$

$$\dot{z}_C = h_C(X,Y,z_D,z_C,\lambda) \tag{6.25}$$

where $F(\cdot)$ in (6.22) describes the dynamics of synchronous machines, the excitation systems, the prime mover and speed governors, and $G(\cdot)$ in (6.23) represents the system network functions. (6.22) and (6.23) involve the transient state variables X and algebraic variables Y respectively. The variable Y usually relates to network bus voltage magnitudes and angles. The long-term dynamics are captured by discrete and/or continuous-time time variables in (6.24) & (6.25), respectively. z_D relates to discrete controls such as tap changers, z_C represents continuous load recovery dynamics. Finally, λ in (6.22) ~ (6.25) denotes changes in demand and the corresponding generation rescheduling. For QSS simulation, (6.22) will be replaced by its equilibrium equation.

Load model plays an important role in voltage stability analysis. Generally, the steady state load characteristics are:

$$P_l = P_S = P_{l0}(\frac{V}{V_0})^{\alpha_s} \qquad (6.26)$$

$$Q_l = Q_S = Q_{l0}(\frac{V}{V_0})^{\beta_s} \qquad (6.27)$$

Where, P_{lo} and Q_{lo} are the powers absorbed by the load at the nominal voltage V_0. And α_S and β_S are the steady-state load exponents.

The tap changing logic at time instant t_k [24] is given as follows:

$$r_{k+1} = \begin{cases} r_k + \Delta r & \text{if } V_2 > V_2^0 + d \quad \text{and} \quad r_k < r_{max} \\ r_k - \Delta r & \text{if } V_2 < V_2^0 - d \quad \text{and} \quad r_k > r_{min} \\ r_k & \text{otherwise} \end{cases} \qquad (6.28)$$

where V_2 is the controlled voltage after OLTC, V_2^0 is the reference voltage, d is half the OLTC dead-band, and r_{max} and r_{min} are the upper and lower tap limits.

6.5.2 Steps involved in QSS Method [24]

The basic QSS simulation methodology is outlined in Fig.6.7. Point A is the equilibrium before the disturbance. Point A' is the equilibrium of these equations after the disturbance. The continuous change from A' to B results from the evolution of λ. The transition from B to B' is from the discrete change of z_D. Also, the load increase scenario (i.e., $\lambda(t)$ function) is known.

(1) Point $A(X_0^-,Y_0^-)$ is an equilibrium of Eqs. 6.23 and 6.22 with $\dot{X}=0$ before the disturbance.

(2) Point $A'(X_0^+,Y_0^+)$ is an equilibrium of Eqs.6.23 and 6.22 with $\dot{X}=0$ after the disturbance.

(3) At $A'(X_0^+,Y_0^+)$, predict the next transition time s that is the shortest internal delays among the active OLTCs.

(4) The continuous change from A' to B results from the evolution of λ and z_C. (Time integration may be required.)

(5) The transition from $B(X_1^-,Y_1^-)$ to $B'(X_1^+,Y_1^+)$ is from the discrete change of z_D.

With initial guess (X_1^-,Y_1^-), the solution of (X_1^+,Y_1^+) can be obtained by solving the following equation:

$$\underbrace{\begin{bmatrix} \dfrac{\partial F_i}{\partial X_j} & \dfrac{\partial F_i}{\partial Y_j} \\[2mm] \dfrac{\partial G_i}{\partial X_j} & \dfrac{\partial G_i}{\partial Y_j} \end{bmatrix}}_{J_{XY}} \begin{bmatrix} X_1^{+(j)} - X_1^{+(j-1)} \\[2mm] Y_1^{+(j)} - Y_1^{+(j-1)} \end{bmatrix} = - \begin{bmatrix} F(X_1^{+(j-1)},Y_1^{+(j-1)}),z_{D1},z_{C1},\lambda(t_1) \\[2mm] G(X_1^{+(j-1)},Y_1^{+(j-1)}),z_{D1},z_{C1},\lambda(t_1) \end{bmatrix}$$

Where j is the iteration count.

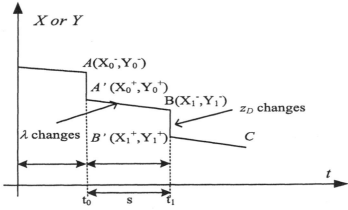

Fig. 6.7.   Time evolution of an X and Y components [24]

To evaluate the system stability condition, reference [24] used the sensitivity of the total reactive power generation Q_g to the various reactive demands Q_i^0:

$$\frac{\partial Q_g}{\partial Q^0} > 0 \text{ indicates stable condition;} \quad \frac{\partial Q_g}{\partial Q^0} < 0 \text{ indicates unstable condi-}$$

tion.

6.5.3 Implementation of the Continuation Method in QSS

Continuation-based Quasi-Steady-State simulation (CQSS) [28] deals with the QSS using the continuation method which is already introduced in previous chapters.

As mentioned before, QSS simulation has been widely used to speed up the long-term voltage stability calculations, which filters out the short-term transients. It deals with the long-term subsystem of the *DAEs* based on the assumption that the transient subsystem is infinitely fast and can be replaced by its equilibrium equations. Therefore in the QSS analysis, the short-term dynamics *dX/dt* in (6.22) is replaced with zero to obtain equilibrium points:

$$0 = F(X, Y, z_D, z_C, \lambda) \tag{6.29}$$

CQSS solves both (6.29) and (6.23), with changing λ, z_D. In CQSS, λ is parameterized to simulate the load restoration.

If you considered the variation of λ, then the continuation method traces the equilibrium defined by (6.29) and (6.23) for a fixed z_D. The Jacobian matrix only involves the derivatives of *F/G* with respect to *X/Y*.

6.5.4 Consideration of Load Change with respect to Time

Similar to the original QSS, there is no time integration involved in CQSS analysis. However the time information is indirectly obtained during the equilibrium tracing. This time information is based on internal delays of discrete controls. The transition time step $s = t_{k+1} - t_k$ is determined by the shortest internal delay or sampling period of the long-term components (z_D). More specifically, in this time step, z_D will be updated according to some control logic. Continuation method introduces parameter λ to easily trace the equilibrium of the system under the step size control. During the process of the OLTC action, λ also varies according to its time-characteristics. Its time function is indicated as $\lambda(t)$ in the CQSS simulation. An approach should be found to appropriately consider how λ changes in the determined time interval $t_{k+1} - t_k$, so as to meet its time function.

In continuation method, from the starting point $A'(X_0^+, Y_0^+)$ as in Fig.7, the prediction and correction steps are given below:

$$-\begin{pmatrix} F_X & F_Y & F_\lambda \\ G_X & G_Y & G_\lambda \\ & e_k^T & \end{pmatrix}\begin{pmatrix} dX \\ dY \\ d\lambda \end{pmatrix} = \begin{pmatrix} 0 \\ 0 \\ \pm 1 \end{pmatrix} \tag{6.30}$$

$$-\begin{pmatrix} F_X & F_Y & F_\lambda \\ G_X & G_Y & G_\lambda \\ & e_k^T & \end{pmatrix}\begin{pmatrix} \Delta X \\ \Delta Y \\ \Delta \lambda \end{pmatrix} = \begin{pmatrix} F \\ G \\ 0 \end{pmatrix} \tag{6.31}$$

From (6.30) and (6.31), we can obtain the point $B(X_1^-, Y_1^-)$ through equations (6.32) and (6.33):

$$\left(X_{k+1}^{-pre} \quad Y_{k+1}^{-pre} \quad \lambda_{k+1}\right)^T = \left(X_k^+ \quad Y_k^+ \quad \lambda_k\right)^T + \sigma_k\left[dX \quad dY \quad d\lambda\right]^T \tag{6.32}$$

$$\left(X_{k+1}^- \quad Y_{k+1}^- \quad \lambda_{k+1}\right)^T = \left(X_{k+1}^{-pre} \quad Y_{k+1}^{-pre} \quad \lambda_{k+1}\right)^T + \left[\Delta X \quad \Delta Y \quad \Delta \lambda\right] \tag{6.33}$$

Where the step size σ_k is:

$$\sigma_k = \frac{\lambda_{k+1} - \lambda_k - \Delta\lambda}{d\lambda} \tag{6.34}$$

In (6.34), λ_{k+1} and λ_k should be known from its time function. If λ is taken as continuation parameter, then $\Delta\lambda$ will be equal to zero in the correction stage. Then the step size σ_k can be obtained as follows:

$$\sigma_k = \frac{\lambda_{k+1} - \lambda_k}{d\lambda} \qquad (6.35)$$

At each time step, this step size should be re-calculated to fit the load variation in this period. It is noted that the Jacobian used in corrector stage depends on the state variables from the predictor stage. However, in order to get $d\lambda$ and $\Delta\lambda$ for the computation of the step size, at first, the same Jacobian as in predictor is also used in corrector. After the approximate step size is obtained, we update the Jacobian in the corrector, get the new $\Delta\lambda$ by solving (6.31) and calculate the step size again by using (6.34). This procedure will be repeated until the error between the updated $\Delta\lambda$ and the old one is within some tolerance.

Note: Null $d\lambda$ detects the singularity of J_{XY}.

The continuous variables z_C, can also be considered in two ways: (i) an explicit integration scheme can be used to update z_C (ii) Represent the continuous load dynamics by the continuous change of load exponent. References [29] and [30] solve the load restoration analytically for a step change in voltage. Based on this derivation, another way to consider the load restoration in the CQSS simulation can also be developed [31].

6.5.5 Numerical Results

6.5.5.1 2-bus system

The 2-bus system has the same parameters as before. For this example the tap position is fixed. The load is increased by 0.5% per second until 10s to simulate the load restoration behavior. The λ value can be directly related to time. The result is shown in Fig.6.8. For comparison, the decoupled time domain simulation result is also superimposed. We can see that as load stops increasing around ten seconds the time domain solution converges to the equilibrium which is identical to that of CQSS.

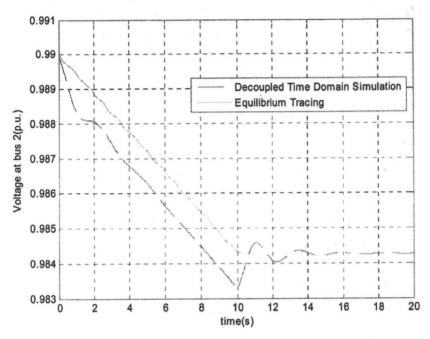

Fig.6.8. Simulation results after 10 seconds of slow load increments

6.5.5.2 CQSS Simulation for New England 39-bus system

The New England 39-bus system is employed here to demonstrate the efficacy of the CQSS method for analysis of long-term voltage instability. The main features of modeling and modification of the system in this simulation are as follows:

- Each generator is represented with the two-axis dynamic model [33] at its equilibrium condition. It includes automatic voltage regulator (AVR), the speed governor, the field and armature current limits as well as the real power generation limit;

The major long-term dynamic phenomena that will be taken into account are: (i) load restoration by on-load tap changer transformers (OLTCs) with fixed load parameter λ (ii) load restoration simulation by directly relating the load parameter λ to time with fixed taps.

Case 1: Load restoration through OLTCs with fixed λ

- 3 OLTCs are equipped at selected load buses (3-2, 16-19, 26-29)
  - Before full load recovery, all the loads at the distribution sides of the OLTCS are constant impedance.
  - Each OLTC has an operating range of 0.8-1.1; the step size of the tap ratio is set as 0.00625; and the time delay is 5 seconds.

The contingency simulated in this case is outage of the line between bus 4 and 5. It is applied to the base case mentioned above. Due to the voltage dependent short-term load characteristic, the system total real power load is 6,238 MW right after the contingency.

During the load restoration: In this step, only tap dynamics are considered for restoring the system load to the pre-contingency level. The OLTCs begin their actions 5s after applying the contingency. The voltages at the distribution side buses of the OLTCs increase whereas voltages at the high-voltage buses decrease due to the load restoration. Fig.6.9 shows the change in voltage at bus 2 with respect to time. Note that this bus is selected from the high-voltage side.

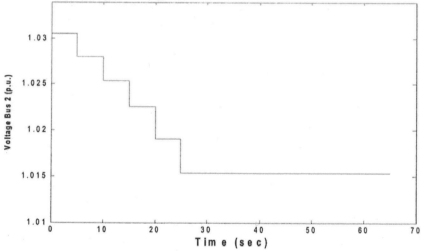

Fig.6.9 Voltage change vs. time during the load restoration

Case 2: Load increment with respect to time with fixed tap position

For this system, to relate the load parameter to time, the OLTCs are fixed now. Constant power loads are used. All the loads are increased by 0.5%

per second simultaneously. The system will collapse after about 53 seconds as shown in Fig.6.10.

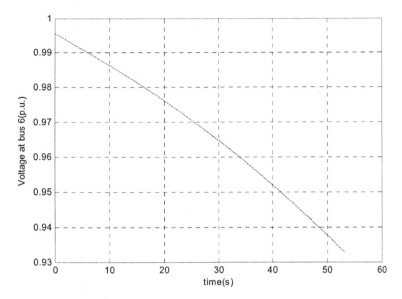

Fig.6.10 Load change (λ) with respect to time

In the scenario, QSS simulation is very fast if one avoids refreshing the Jacobian matrix. However, the Jacobian has to be updated when some of the devices are hitting their limits. Although CQSS involves two Jacobian matrices in the predictor and the corrector, they could be the same for fast computation. Furthermore, the predictor is much less computationally demanding than the corrector. For the predictor we have to solve only once the equation (30). Corrector needs iterative solution similar to solving the equations in the original QSS simulation. Thus, the simulation time in CQSS is almost the same as QSS under light load conditions.

References

[1] Astic, J. Y., Bihain, A., Jerosolimski, M., "The mixed Adams-BDF variable step size algorithm to simulate transient and long-term phenomena in power systems", *IEEE Transactions on Power Systems*, vol. 9, pp.929-935, May 1994.

[2] Iavernaro, F., Scala, M. L., Mazzia, F., "Boundary values methods for time-domain simulation of power system dynamic behavior", IEEE Transactions on Circuits and Systems I: Fundamental Theory and Applications, vol. 45, pp.50-63, Jan 1998.

[3] Liu, C.-W., Thorp, James S., "New methods for computing power system dynamic response for real-time transient stability prediction", IEEE Transactions on Circuits and Systems I: Fundamental Theory and Applications, vol. 47, pp.324-337, March 2000.

[4] Fankhauser, H., Aneros, K., Edris, A., Torseng, S., "Advanced simulation techniques for the analysis of power system dynamics", IEEE Computer Applications in Power, vol. 3, pp.31-36, Oct 1990.

[5] Kurita, A., Okubo, H., Oki, K., Agematsu, S., Klapper, D.B., Miller, N. W., Price, W.W., Sanchez-Gasca, J.J., Wirgau, K.A., Younkins, T.D., "Multiple time-scale power system dynamic simulation", IEEE Transactions on Power Systems, vol. 8, pp.216-223, Feb 1993.

[6] Stubbe, M., Bihain, A., Deuse, J., Baader, J.C., "STAG-a new unified software program for the study of the dynamic behaviour of electrical power systems", IEEE Transactions on Power Systems, vol. 4, pp.129-138, Feb 1989.

[7] Stubbe, M and Deuse, J," Dynamic Simulation of Voltage Collapses" IEEE Transactions on Power Systems, Vol.8, No.3, pp.894-904, August 1993

[8] Vernotte, P., Panciatici, B., Myers, J.P., Antoine, J., Deuse, J., and Stubbe, M., "High fidelity simulation power system dynamics," *IEEE Comput. Appl. Power*, vol. 8, no. 1, pp. 37–41, Jan. 1995

[9] Sanchez-Gasca, J.J., R. D'Aquila, Paserba, J.J., Price, W.W., Klapper, D.B., and Hu, I.P., "Extended-term dynamic simulation using variable time step integration," *IEEE Comput. Appl. Power*, vol. 6, no. 4, pp. 23–28, Oct. 1993.

[10] Zhou, Y., Ajjarapu, V., "A Novel Approach to Trace Time Domain Trajectories of Power Systems in Multiple Time Scales, IEEE Transactions On Power Systems, Vol.20, No. 1, 149-155, February 2005.

[11] Ascher, U.M., Petzold, Linda R., Computer methods for ordinary differential equations and differential-algebraic equations, Philadelphia: Society for Industrial and Applied Mathematics, 1998.

[12] Hairer, E., Wanner, G., Solving ordinary differential equations II: stiff and differential-algebraic problems, Springer-Verlag, 1991.

[13] Lambert, J. D., Numerical methods for ordinary differential systems: the initial value problems, Wiley, 1991.

[14] Iserles, A., A first course in the numerical analysis of differential equations, Cambridge University Press, 1996.

[15] Deuflhard, P., Bornemann, F., Scientific computing with ordinary differential equations, Springer, 2002.

[16] Butcher, J.C., "Numerical methods for ordinary differential equations in the 20th century", Numerical analysis: historical developments in the 20th century, edited by C. Brezinski, L. Wuytack, Elsevier, 2001

[17] Janovsky, V., Liberda, O., "Continuation of invariant subspaces via the recursive projection method", Applications of Mathematics, vol.48, pp.241-255, 2003.

[18] Shroff, G.M., Keller, H.B., "Stabilization of unstable procedures: the recursive projection method", SIAM Journal on Numerical Analysis, vol.30, pp.1099-1120, August, 1993.

[19] Sauer, P. W., Pai, M. A., Power System Dynamics and Stability, Prentice-Hall Inc, 1998.

[20] Feng, Z., Ajjarapu, V., Long, B., "Identification of voltage collapse through direct equilibrium tracing", IEEE Transactions on Power Systems, vol.15, pp.342-349, February 2000.

[21] Golub, G. H., Loan, C. V., Matrix Computations, Johns Hopkins University Press, Baltimore, MD, 1983

[22] Yang, D., Ajjarapu, V., "A decoupled time-domain simulation method via invariant subspace partition for power system analysis," *IEEE Transactions on Power Systems*, vol.21, no.1, February 2006

[23] Van Cutsem, T., Vournas, C., *Voltage Stability of Electric Power Systems*, Kluwer Academic Publishers, Norwell, MA, 1998.

[24] Van Cutsem, T., Jacquemart, Y., Marquet, J., Pruvot, P., "A comprehensive analysis of mid-term voltage stability," *IEEE Trans. on Power Systems*, vol. 10, no. 2, pp. 1173-1182, May 1995.

[25] Van Cutsem, T., "An approach to corrective control of voltage instability using simulation and sensitivity," *IEEE Trans. on Power Systems*, vol. 10, no.2, pp. 616-622, May 1995.

[26] Van Cutsem, T., Mailhot, R., "Validation of a fast voltage stability analysis method on the hydro-quebec system," *IEEE Trans. on Power Systems*, vol. 12, no.1, pp. 282-292, Feb. 1997.

[27] Van Cutsem, T., Vournas, C.D., "Voltage stability analysis in transient and mid-term time scales," *IEEE Trans. on Power Systems*, vol. 11, no.1, pp. 146-154, Feb. 1996.

[28] Wang, Q., Song, H., Ajjarapu, V., "Continuation Based Quasi-Steady-State Analysis," *IEEE Transactions on Power Systems*, vol.21, no.1, February 2006

[29] Karlsson, D., Hill, D. J., "Modeling and identification of nonlinear dynamic loads in power systems," IEEE Trans. on Power Systems, vol. 9, no. 1, Feb. 1994, pp. 157-166.

[30] Hill, D. J., "Nonlinear dynamic load models with recovery for voltage stability studies," IEEE Trans. on Power Systems, vol. 8, no. 1, Feb. 1993, pp. 166-172.

[31] Wang, Q., Ajjarapu, V., "A novel approach to implement generic load restoration in continuation-based quasi-steady-state analysis," *IEEE PES Technical Letters*, PESL-00076-2002, 17 May 2004 - 1 July 2004.

Appendix

A. Data of 2-bus test system

A1. One line diagram

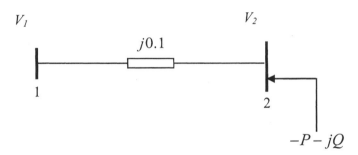

A2. The IEEE format: Base case power flow data of the 2-bus system

BUS DATA

```
1 01alpha    1  0  3 1.0000    0.00       0.00      0.00    210.00    0.00  100.00
1.0000    0.00     0.00  0.0000   0.0000     0    1
2 01kappa    1  0  0 1.0000    0.00    140.00      0.00      0.00    0.00  100.00
0.0000    0.00     0.00  0.0000   0.0000     0    2
-999
```

BRANCH DATA
```
1    2  1  0 1 0  0.000000   0.100000    0.00000   0.    0.    0.    0 0  0.0000
0.0  0.0     0.0   0.0       0.0     0.0        1
-999
```

A3. The dynamic data of the 2-bus system

X_{d1}	X_{q1}	X'_{d1}	X'_{q1}	R_{s1}	T'_{d01}	T'_{q01}
1.67	1	0.232	0.466	0.0002	5.4	0.88
M_1	D_1	K_{e1}	T_{e1}	S_{e1}	K_{a1}	T_{a1}
52	5	1	0.79	0	30	0.02
K_{f1}	T_{f1}	T_{ch1}	T_{g1}	R_1	ω_{ref}	
0.03	1	9.79	0.12	0.05	1	

B. Data of New England test system

B1. One line diagram

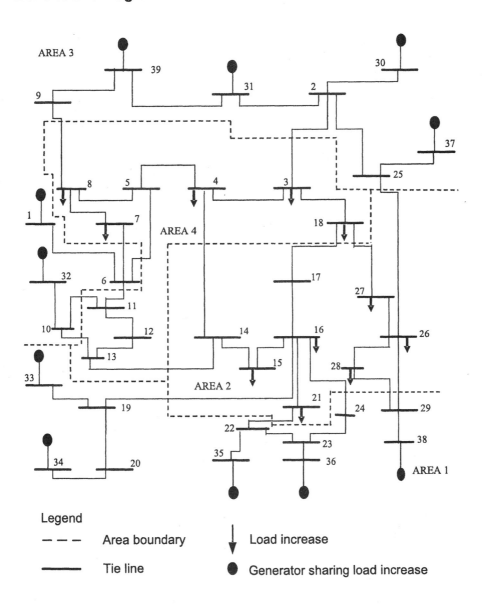

Legend

— — —   Area boundary

↓   Load increase

———   Tie line

●   Generator sharing load increase

B2. The IEEE format: Base case power flow data of the New England system

BUS DATA
```
    1 BUS1          1  1  0 1.0436 -13.39    0.00       0.00      0.00     0.00
0.00 0.0000   0.0000   0.0000   0.0000   0.0000    0    1
    2 BUS2          1  1  0 1.0378 -11.21    0.00       0.00      0.00     0.00
0.00 0.0000   0.0000   0.0000   0.0000   0.0000    0    2
    3 BUS3          1  1  0 1.0056 -13.87  322.00     122.40      0.00     0.00
0.00 0.0000   0.0000   0.0000   0.0000   0.0000    0    3
    4 BUS4          1  1  0 0.9864 -14.01  500.00     184.00      0.00     0.00
0.00 0.0000   0.0000   0.0000   0.0000   0.0000    0    4
    5 BUS5          1  1  0 0.9924 -12.24    0.00       0.00      0.00     0.00
0.00 0.0000   0.0000   0.0000   0.0000   0.0000    0    5
    6 BUS6          1  1  0 0.9956 -11.40    0.00       0.00      0.00     0.00
0.00 0.0000   0.0000   0.0000   0.0000   0.0000    0    6
    7 BUS7          1  1  0 0.9851 -13.75  233.80      84.00      0.00     0.00
0.00 0.0000   0.0000   0.0000   0.0000   0.0000    0    7
    8 BUS8          1  1  0 0.9843 -14.32  522.00     176.00      0.00     0.00
0.00 0.0000   0.0000   0.0000   0.0000   0.0000    0    8
    9 BUS9          1  1  0 1.0233 -14.58    0.00       0.00      0.00     0.00
0.00 0.0000   0.0000   0.0000   0.0000   0.0000    0    9
   10 BUS10         1  1  0 1.0060  -9.41    0.00       0.00      0.00     0.00
0.00 0.0000   0.0000   0.0000   0.0000   0.0000    0   10
   11 BUS11         1  1  0 1.0013 -10.10    0.00       0.00      0.00     0.00
0.00 0.0000   0.0000   0.0000   0.0000   0.0000    0   11
   12 BUS12         1  1  0 0.9876 -10.23    8.50      88.00      0.00     0.00
0.00 0.0000   0.0000   0.0000   0.0000   0.0000    0   12
   13 BUS13         1  1  0 1.0014 -10.23    0.00       0.00      0.00     0.00
0.00 0.0000   0.0000   0.0000   0.0000   0.0000    0   13
   14 BUS14         1  1  0 0.9947 -12.18    0.00       0.00      0.00     0.00
0.00 0.0000   0.0000   0.0000   0.0000   0.0000    0   14
   15 BUS15         1  1  0 0.9909 -13.33  320.00     153.00      0.00     0.00
0.00 0.0000   0.0000   0.0000   0.0000   0.0000    0   15
   16 BUS16         1  .1  0 1.0043 -12.16  329.40     132.30      0.00     0.00
0.00 0.0000   0.0000   0.0000   0.0000   0.0000    0   16
   17 BUS17         1  1  0 1.0076 -13.11    0.00       0.00      0.00     0.00
0.00 0.0000   0.0000   0.0000   0.0000   0.0000    0   17
   18 BUS18         1  1  0 1.0055 -13.85  158.00      30.00      0.00     0.00
0.00 0.0000   0.0000   0.0000   0.0000   0.0000    0   18
   19 BUS19         1  1  0 1.0432  -7.90    0.00       0.00      0.00     0.00
0.00 0.0000   0.0000   0.0000   0.0000   0.0000    0   19
   20 BUS20         1  1  0 0.9938  -9.51  680.00     103.00      0.00     0.00
0.00 0.0000   0.0000   0.0000   0.0000   0.0000    0   20
   21 BUS21         1  1  0 1.0122  -9.83  274.00     115.00      0.00     0.00
0.00 0.0000   0.0000   0.0000   0.0000   0.0000    0   21
   22 BUS22         1  1  0 1.0387  -5.44    0.00       0.00      0.00     0.00
0.00 0.0000   0.0000   0.0000   0.0000   0.0000    0   22
```

```
   23 BUS23          1  1  0 1.0322   -5.65   247.50     84.60      0.00      0.00
0.00 0.0000   0.0000    0.0000   0.0000   0.0000     0    23
   24 BUS24          1  1  0 1.0029 -12.07   308.60     92.20      0.00      0.00
0.00 0.0000   0.0000    0.0000   0.0000   0.0000     0    24
   25 BUS25          1  1  0 1.0461 -10.01   224.00     47.20      0.00      0.00
0.00 0.0000   0.0000    0.0000   0.0000   0.0000     0    25
   26 BUS26          1  1  0 1.0299 -11.38   139.00     47.00      0.00      0.00
0.00 0.0000   0.0000    0.0000   0.0000   0.0000     0    26
   27 BUS27          1  1  0 1.0136 -13.39   281.00     75.50      0.00      0.00
0.00 0.0000   0.0000    0.0000   0.0000   0.0000     0    27
   28 BUS28          1  1  0 1.0308   -8.00   206.00     27.60      0.00      0.00
0.00 0.0000   0.0000    0.0000   0.0000   0.0000     0    28
   29 BUS29          1  1  0 1.0318   -5.22   283.50    126.90      0.00      0.00
0.00 0.0000   0.0000    0.0000   0.0000   0.0000     0    29
   30 BUS30          1  1  2 1.0475   -8.96     0.00      0.00    230.00    206.87
0.00 1.0475 380.00 -100.00   0.0000   0.0000     0    30
   31 BUS31          1  1  3 0.9820    0.00     0.00      0.00    722.53    274.61
0.00 0.9820 600.00 -300.00   0.0000   0.0000     0    31
   32 BUS32          1  1  2 0.9831   -1.58     0.00      0.00    630.00    254.00
0.00 0.9831 500.00 -300.00   0.0000   0.0000     0    32
   33 BUS33          1  1  2 0.9972   -2.84     0.00      0.00    612.00    152.86
0.00 0.9972 500.00 -300.00   0.0000   0.0000     0    33
   34 BUS34          1  1  2 1.0123   -4.50     0.00      0.00    488.00    236.74
0.00 1.0123 450.00 -250.00   0.0000   0.0000     0    34
   35 BUS35          1  1  2 1.0493   -0.58     0.00      0.00    630.00    290.62
0.00 1.0493 600.00 -250.00   0.0000   0.0000     0    35
   36 BUS36          1  1  2 1.0635    2.00     0.00      0.00    540.00    148.33
0.00 1.0635 500.00 -220.00   0.0000   0.0000     0    36
   37 BUS37          1  1  2 1.0278   -3.42     0.00      0.00    520.00     48.40
0.00 1.0278 500.00 -220.00   0.0000   0.0000     0    37
   38 BUS38          1  1  2 1.0265    1.74     0.00      0.00    810.00    138.33
0.00 1.0265 500.00 -300.00   0.0000   0.0000     0    38
   39 BUS39          1  1  2 1.0300 -14.68  1104.00    250.00   1000.00    123.30
0.00 1.0300 900.00 -800.00   0.0000   0.0000     0    39
-999
```

BRANCH DATA

```
    1    2  1  1 1 0  0.003500  0.041100   0.69870     0.    0.    0.    0 0
0.0000     0.00 0.0000 0.0000 0.0000   0.0000 0.0000     1
    1   39  1  1 1 0  0.002000  0.050000   0.37500     0.    0.    0.    0 0
0.0000     0.00 0.0000 0.0000 0.0000   0.0000 0.0000     2
    1   39  1  1 2 0  0.002000  0.050000   0.37500     0.    0.    0.    0 0
0.0000     0.00 0.0000 0.0000 0.0000   0.0000 0.0000     3
    2    3  1  1 1 0  0.001300  0.015100   0.25720     0.    0.    0.    0 0
0.0000     0.00 0.0000 0.0000 0.0000   0.0000 0.0000     4
    2   25  1  1 1 0  0.007000  0.008600   0.14600     0.    0.    0.    0 0
0.0000     0.00 0.0000 0.0000 0.0000   0.0000 0.0000     5
    3    4  1  1 1 0  0.001300  0.021300   0.22140     0.    0.    0.    0 0
0.0000     0.00 0.0000 0.0000 0.0000   0.0000 0.0000     6
    3   18  1  1 1 0  0.001100  0.013300   0.21380     0.    0.    0.    0 0
0.0000     0.00 0.0000 0.0000 0.0000   0.0000 0.0000     7
```

```
   4    5  1  1 1 0   0.000800   0.012800   0.13420    0.     0.     0.     0 0
0.0000        0.00 0.0000 0.0000 0.0000   0.0000 0.0000         8
   4   14  1  1 1 0   0.000800   0.012900   0.13820    0.     0.     0.     0 0
0.0000        0.00 0.0000 0.0000 0.0000   0.0000 0.0000         9
   5    6  1  1 1 0   0.000200   0.002600   0.04340    0.     0.     0.     0 0
0.0000        0.00 0.0000 0.0000 0.0000   0.0000 0.0000        10
   5    8  1  1 1 0   0.000800   0.011200   0.14760    0.     0.     0.     0 0
0.0000        0.00 0.0000 0.0000 0.0000   0.0000 0.0000        11
   6    7  1  1 1 0   0.000600   0.009200   0.11300    0.     0.     0.     0 0
0.0000        0.00 0.0000 0.0000 0.0000   0.0000 0.0000        12
   6   11  1  1 1 0   0.000700   0.008200   0.13890    0.     0.     0.     0 0
0.0000        0.00 0.0000 0.0000 0.0000   0.0000 0.0000        13
   7    8  1  1 1 0   0.000400   0.004600   0.07800    0.     0.     0.     0 0
0.0000        0.00 0.0000 0.0000 0.0000   0.0000 0.0000        14
   8    9  1  1 1 0   0.002300   0.036300   0.38040    0.     0.     0.     0 0
0.0000        0.00 0.0000 0.0000 0.0000   0.0000 0.0000        15
   9   39  1  1 1 0   0.001000   0.025000   1.20000    0.     0.     0.     0 0
0.0000        0.00 0.0000 0.0000 0.0000   0.0000 0.0000        16
  10   11  1  1 1 0   0.000400   0.004300   0.07290    0.     0.     0.     0 0
0.0000        0.00 0.0000 0.0000 0.0000   0.0000 0.0000        17
  10   13  1  1 1 0   0.000400   0.004300   0.07290    0.     0.     0.     0 0
0.0000        0.00 0.0000 0.0000 0.0000   0.0000 0.0000        18
  13   14  1  1 1 0   0.000900   0.010100   0.17230    0.     0.     0.     0 0
0.0000        0.00 0.0000 0.0000 0.0000   0.0000 0.0000        19
  14   15  1  1 1 0   0.001800   0.021700   0.36600    0.     0.     0.     0 0
0.0000        0.00 0.0000 0.0000 0.0000   0.0000 0.0000        20
  15   16  1  1 1 0   0.000900   0.009400   0.17100    0.     0.     0.     0 0
0.0000        0.00 0.0000 0.0000 0.0000   0.0000 0.0000        21
  16   17  1  1 1 0   0.000700   0.008900   0.13420    0.     0.     0.     0 0
0.0000        0.00 0.0000 0.0000 0.0000   0.0000 0.0000        22
  16   19  1  1 1 0   0.001600   0.019500   0.30400    0.     0.     0.     0 0
0.0000        0.00 0.0000 0.0000 0.0000   0.0000 0.0000        23
  16   21  1  1 1 0   0.000800   0.013500   0.25480    0.     0.     0.     0 0
0.0000        0.00 0.0000 0.0000 0.0000   0.0000 0.0000        24
  16   24  1  1 1 0   0.000300   0.005900   0.06800    0.     0.     0.     0 0
0.0000        0.00 0.0000 0.0000 0.0000   0.0000 0.0000        25
  17   18  1  1 1 0   0.000700   0.008200   0.13190    0.     0.     0.     0 0
0.0000        0.00 0.0000 0.0000 0.0000   0.0000 0.0000        26
  17   27  1  1 1 0   0.001300   0.017300   0.32160    0.     0.     0.     0 0
0.0000        0.00 0.0000 0.0000 0.0000   0.0000 0.0000        27
  21   22  1  1 1 0   0.000800   0.014000   0.25650    0.     0.     0.     0 0
0.0000        0.00 0.0000 0.0000 0.0000   0.0000 0.0000        28
  22   23  1  1 1 0   0.000600   0.009600   0.18460    0.     0.     0.     0 0
0.0000        0.00 0.0000 0.0000 0.0000   0.0000 0.0000        29
  23   24  1  1 1 0   0.002200   0.035000   0.36100    0.     0.     0.     0 0
0.0000        0.00 0.0000 0.0000 0.0000   0.0000 0.0000        30
  25   26  1  1 1 0   0.003200   0.032300   0.51300    0.     0.     0.     0 0
0.0000        0.00 0.0000 0.0000 0.0000   0.0000 0.0000        31
  26   27  1  1 1 0   0.001400   0.014700   0.23960    0.     0.     0.     0 0
0.0000        0.00 0.0000 0.0000 0.0000   0.0000 0.0000        32
```

```
 26    28   1   1 1 0   0.004300   0.047400   0.78020    0.    0.    0.    0 0
0.0000      0.00 0.0000 0.0000 0.0000   0.0000 0.0000    33
 26    29   1   1 1 0   0.005700   0.062500   1.02900    0.    0.    0.    0 0
0.0000      0.00 0.0000 0.0000 0.0000   0.0000 0.0000    34
 28    29   1   1 1 0   0.001400   0.015100   0.24900    0.    0.    0.    0 0
0.0000      0.00 0.0000 0.0000 0.0000   0.0000 0.0000    35
  2    30   1   1 1 1   0.000000   0.018100   0.00000    0.    0.    0.    0 0
1.0250      0.00 0.0000 0.0000 0.0000   0.0000 0.0000    36
  6    31   1   1 1 1   0.000000   0.050000   0.00000    0.    0.    0.    0 0
1.0700      0.00 0.0000 0.0000 0.0000   0.0000 0.0000    37
  6    31   1   1 2 1   0.000000   0.050000   0.00000    0.    0.    0.    0 0
1.0700      0.00 0.0000 0.0000 0.0000   0.0000 0.0000    38
 10    32   1   1 1 1   0.000000   0.020000   0.00000    0.    0.    0.    0 0
1.0700      0.00 0.0000 0.0000 0.0000   0.0000 0.0000    39
 12    11   1   1 1 1   0.001600   0.043500   0.00000    0.    0.    0.    0 0
1.0060      0.00 0.9200 1.0800 0.0000   0.9500 1.0500    40
 12    13   1   1 1 1   0.001600   0.043500   0.00000    0.    0.    0.    0 0
1.0060      0.00 0.9200 1.0800 0.0000   0.9500 1.0500    41
 19    20   1   1 1 1   0.000700   0.013800   0.00000    0.    0.    0.    0 0
1.0600      0.00 0.9200 1.0800 0.0000   0.9500 1.0500    42
 19    33   1   1 1 1   0.000700   0.014200   0.00000    0.    0.    0.    0 0
1.0700      0.00 0.0000 0.0000 0.0000   0.0000 0.0000    43
 20    34   1   1 1 1   0.000900   0.018000   0.00000    0.    0.    0.    0 0
1.0250      0.00 0.8750 1.1250 0.0000   0.9500 1.0500    44
 22    35   1   1 1 1   0.000000   0.014300   0.00000    0.    0.    0.    0 0
1.0250      0.00 0.0000 0.0000 0.0000   0.0000 0.0000    45
 23    36   1   1 1 1   0.000500   0.027200   0.00000    0.    0.    0.    0 0
1.0000      0.00 0.0000 0.0000 0.0000   0.0000 0.0000    46
 25    37   1   1 1 1   0.000600   0.023200   0.00000    0.    0.    0.    0 0
1.0250      0.00 0.0000 0.0000 0.0000   0.0000 0.0000    47
 29    38   1   1 1 1   0.000800   0.015600   0.00000    0.    0.    0.    0 0
1.0250      0.00 0.0000 0.0000 0.0000   0.0000 0.0000    48
-999
LOSS ZONES
-99
INTERCHANGE DATA FOLLOWS
-9
TIE LINES FOLLOW
-999
```

B3. The Dynamic Data of the New England System

NEW_ENGLAND SYSTEM STABILITY RELATED PARAMETERS OF GENERATOR & EXCITATION & GOVERNOR & SVC & OLTC & DYNAMIC LOADS

Generator transient parameter follows

```
          1         2         3         4         5         6         7
               9        10        11
234567890123456789012345678901234567890123456789012345678901234567890123
567890123456789012345678901234567890
```

um Gen_name	Xd	Xq	X'd	X'q	Rs	T'do
'qo Mg	Dg					
30 BUS30	0.1000	0.0690	0.0310	0.0690	0.0002	10.2000
.010 84.000	5.000					
31 BUS31	0.2590	0.2820	0.0700	0.1700	0.0002	6.5600
.5000 60.600	5.000					
32 BUS32	0.2500	0.2370	0.0530	0.0880	0.0002	5.7000
1.5000 71.600	5.000					
33 BUS33	0.2620	0.2580	0.0440	0.1660	0.0002	5.6900
1.5000 57.200	5.000					
34 BUS34	0.6700	0.6200	0.1320	0.1660	0.0002	5.4000
0.4400 52.000	5.000					
35 BUS35	0.2540	0.2410	0.0500	0.0810	0.0002	7.3000
0.4000 69.600	5.000					
36 BUS36	0.2950	0.2920	0.0490	0.1860	0.0002	5.6600
1.5000 52.800	5.000					
37 BUS37	0.2900	0.2800	0.0570	0.0910	0.0010	6.7000
0.4100 48.600	5.000					
38 BUS38	0.2110	0.2050	0.0570	0.0590	0.0002	4.7900
1.9600 69.000	5.000					
39 BUS39	0.0200	0.0190	0.0060	0.0080	0.0002	7.0000
0.7000 1000.000	10.000					
-999						

Generator control system (Exciter + AVR + governor) parameter

Num Gen_name	Ke	Te	Se	Ka	Ta	Kf
Tf Tch	Tg	Rg				
30 BUS30	1.0000	0.2500	0.0000	20.0000	0.0600	0.0400
1.0000 1.6000	0.2000	0.0500				
31 BUS31	1.0000	0.4100	0.0000	40.0000	0.0500	0.0600
0.5000 54.1000	0.4500	0.0500				
32 BUS32	1.0000	0.5000	0.0000	40.0000	0.0600	0.0800
1.0000 10.0000	3.0000	0.0500				
33 BUS33	1.0000	0.5000	0.0000	40.0000	0.0600	0.0800
1.0000 10.1800	0.2400	0.0500				
34 BUS34	1.0000	0.7900	0.0000	30.0000	0.0200	0.0300
1.0000 9.7900	0.1200	0.0500				
35 BUS35	1.0000	0.4700	0.0000	40.0000	0.0200	0.0800
1.2500 10.0000	3.0000	0.0500				
36 BUS36	1.0000	0.7300	0.0000	30.0000	0.0200	0.0300
1.0000 7.6800	0.2000	0.0500				
37 BUS37	1.0000	0.5300	0.0000	40.0000	0.0200	0.0900
1.2600 7.0000	3.0000	0.0500				

```
 38 BUS38           1.0000   1.4000   0.0000  20.0000   0.0200   0.0300
1.0000    6.1000   0.3800   0.0500
 39 BUS39           1.0000   1.0000   0.0000  20.0000   0.0200   0.0300
1.0000   10.0000   2.0000   0.0500
-999
```

Index